Springer Theses

Recognizing Outstanding Ph.D. Research

For further volumes:
http://www.springer.com/series/8790

Aims and Scope

The series "Springer Theses" brings together a selection of the very best Ph.D. theses from around the world and across the physical sciences. Nominated and endorsed by two recognized specialists, each published volume has been selected for its scientific excellence and the high impact of its contents for the pertinent field of research. For greater accessibility to non-specialists, the published versions include an extended introduction, as well as a foreword by the student's supervisor explaining the special relevance of the work for the field. As a whole, the series will provide a valuable resource both for newcomers to the research fields described, and for other scientists seeking detailed background information on special questions. Finally, it provides an accredited documentation of the valuable contributions made by today's younger generation of scientists.

Theses are accepted into the series by invited nomination only and must fulfill all of the following criteria

- They must be written in good English.
- The topic should fall within the confines of Chemistry, Physics and related interdisciplinary fields such as Materials, Nanoscience, Chemical Engineering, Complex Systems and Biophysics.
- The work reported in the thesis must represent a significant scientific advance.
- If the thesis includes previously published material, permission to reproduce this must be gained from the respective copyright holder.
- They must have been examined and passed during the 12 months prior to nomination.
- Each thesis should include a foreword by the supervisor outlining the significance of its content.
- The theses should have a clearly defined structure including an introduction accessible to scientists not expert in that particular field.

Colm Duffy

Heteroaromatic Lipoxin A$_4$ Analogues

Synthesis and Biological Evaluation

Doctoral Thesis accepted by
University College Dublin, Ireland

 Springer

Author
Dr. Colm Duffy
UCD School of Chemistry
  and Chemical Biology
Centre for Synthesis
  and Chemical Biology
University College Dublin
Belfield, Dublin 4
Ireland
e-mail: cduffy@mrc-lmb.cam.ac.uk

Supervisor
Prof. Dr. Pat Guiry
UCD School of Chemistry
  and Chemical Biology
Centre for Synthesis
  and Chemical Biology
University College Dublin
Belfield, Dublin 4
Ireland
e-mail: p.guiry@ucd.ie

ISSN 2190-5053
ISBN 978-3-642-43892-9
DOI 10.1007/978-3-642-24632-6
Springer Heidelberg Dordrecht London New York

e-ISSN 2190-5061
ISBN 978-3-642-24632-6 (eBook)

Springer is part of Springer Science+Business Media (www.springer.com)

Parts of this thesis have been published in the following journal articles:

Singh S, Duffy CD, Shah STA, Guiry PJ (2008) $ZrCl_4$ as an efficient catalyst for a novel one-pot protection/deprotection synthetic methology. J Org Chem 73: 6429

Duffy CD, Maderna P, McCarthy C, Loscher CE, Catherine G, Guiry PJ (2010) Synthesis and biological evaluation of pyridine-containing lipoxin A_4 analogues. Chem Med Chem 5:517

Duffy CD, Guiry PJ (2010) Recent advances in the chemistry and biology of stable synthetic lipoxin analogues. Med Chem Comm 1:249

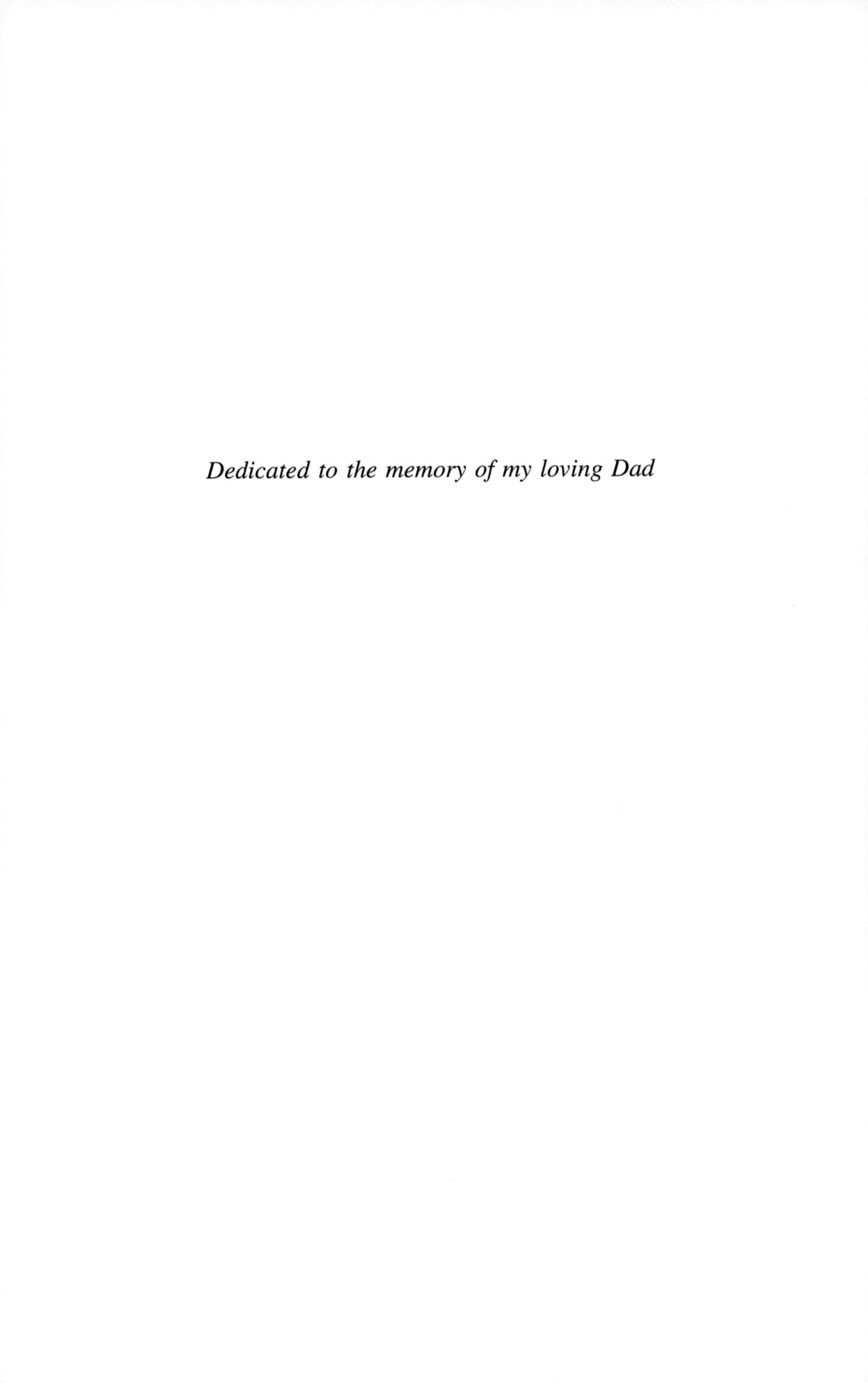

Dedicated to the memory of my loving Dad

Supervisor's Foreword

Both in vivo and in vitro studies have shown that lipoxins, in particular LXA$_4$ isolated in 1984, regulate leukocyte function and inhibit chemotaxis of polymorphonuclear (PMN) leukocytes. Importantly, lipoxins also act as to mediate inflammatory responses by interfering with neutrophil and eosinophil adhesion and migration. As with many natural products, minimal quantities result from isolating these compounds from natural sources. In addition, the accumulation of LXA$_4$ at the site of inflammation is short lived as they are rapidly metabolised. These issues are major obstacles to the application of these compounds as important pharmacological agents. Therefore, it is important to prepare and evaluate the biological activity of a novel range of stable LXA$_4$ analogues that are designed to inhibit, resist, or more slowly undergo metabolism and should therefore have a longer pharmacological activity.

This thesis describes our continuing research programme on synthetic efforts of mimicking the core structure of the native LXA$_4$ by replacing the triene unit, in this study, with a chemically stable heteroaromatic groups (Fig. 1, Approach B).

Dr Duffy has done an excellent job herein by reviewing the chemistry/biology

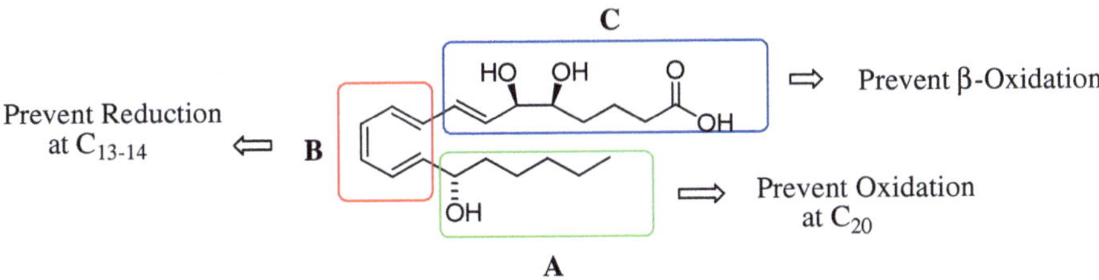

Fig. 1 Targeted domains for modifications of the native LXA$_4$

of stable lipoxin analogues, a field that warranted such a comprehensive collation of recent research efforts. He prepared for the first time ever a pyridine-containing LXA$_4$ analogue in enantiomerically pure form. Biological evaluation determined that both epimers at the benzylic position suppress key cytokines known to be involved in inflammatory disease, with the (R)-epimer proving most efficacious. He also developed an excellent route to a related thiophene-containing analogue that also showed interesting biological activity. Both routes have inspired further work in the research group where we are currently investigating the synthesis of further examples heteroaromatic analogues for biological evaluation.

He played a key role in the development of zirconium tetrachloride as a novel catalyst for a one-pot protection/deprotection methodology, something we are currently exploiting in a different research area within the group—the synthesis of the δ-lactone marine natural products, (+)-tanikolide and (−)-malyngolide.

Belfield, Dublin 4, October 2011   Prof. Dr. Pat Guiry

Acknowledgments

Firstly, I would like to express my sincere gratitude to Prof. Pat Guiry for giving me the opportunity to work as part of his research group. My PhD has been a very positive experience and I am forever grateful for everything you have taught me. It was a pleasure and a privilege to work for you.

I would like to thank University College Dublin for the award of an Ad Astra Scholarship which funded my PhD research.

I have made many friends during my time in UCD. I feel extremely fortunate to have worked within the Guiry group especially with Miriam, Billy and Sean. You all made my workday so enjoyable and I thank you for your kind friendship. I am forever indebted to Dr Surendra Singh for being an excellent mentor to me in the lab. I thank you for all that you taught me. Above all I am grateful that I have absorbed your strong sense of integrity, an essential value when pursuing a scientific career. I wish you the very best in your new position in Delhi and I am confident that you will make an excellent researcher. I would like to say a special thank you to the members of the Lipoxin group. Tasadaque, Gavin and Caroline have been a huge support during my time in the lab. I am very grateful to Christina and Barry for their detailed proofreading of my thesis and their valuable feedback. I would like to thank all the past and present members of the Guiry group that I had the pleasure of working with. I would like to wish you all the very best in your careers and I hope that our paths will cross sometime in the future. I am also forever grateful to Sabrina, Trish, Nello and Caitríona. You offered friendship and support and you will remain close friends. I hope that your futures will be filled with success and happiness.

It was a pleasure to work with the technical and stores staff in UCD. I thank Jimmy Muldoon, Yannick Ortin, Dilip Rai, Dermot Keenan, Kevin Conboy and Adam Coburn for their help with all the analysis. I am also grateful to Gerry Flynn, Mary Flannery and Patrick Waldron for their excellent running of stores.

I thank my transfer assessment panel, Dr Paul Evans and Dr Francesca Paradisi for their help and advice during my time in UCD. I am also grateful to Dr Chris Braddock for taking the time to correct my thesis and also being an excellent external examiner.

To my good friends from Knocklyon. Your friendship and support during this very difficult year will never be forgotten. Paul, Jim, Phil, Dave and Peter, I thank you for your solid support at a time when I needed it so much. I appreciate your willingness to listen whenever I needed to talk. Thanks for putting up with me! To John, Aido, Nick, Becky, Bronagh, Matthew and Darragh, your friendship means so much to me. I hope one day I will be able to repay the kindness you have shown me.

A very big thank you must be said to my family. I am blessed to have such a special Mom. You truly are a magnificent, loving person and I cannot thank you enough for everything you have done and continue to do for me. I would also like to thank my brother Mark, my sister Emma and my Nana for being so supportive during my PhD.

Another special person in my life is Elaine. I thank you for the love and kindness that you have shown me. I would still be writing this thesis if it wasn't for all your help! You were always there for me when I needed you the most. You are an incredible person with a unique ability to inspire others. I love you with all my heart.

Finally, I would like to dedicate this thesis to my loving Dad. You always supported me in everything I did whether it was on the side of a muddy pitch, musically and most especially during my studies. Thank you for the love, guidance and support you so willingly gave me throughout my life. There is not a day that goes by when I don't miss you. You will always be loved and never forgotten.

General Experimental

All reactions were carried out under an inert atmosphere of nitrogen using oven dried glassware and reagents were purchased from Sigma-Aldrich apart from 1,2-dibromotetrafluoroethane which was purchased from Apollo Scientific. Oxygen-free nitrogen was obtained from BOC gases and used without further drying. Diethyl ether, tetrahydrofuran and dichloromethane were obtained from a PureSolv-300-3-MD dry solvent dispenser and used without further purification. Dimethyacetamide was purchased from Sigma-Aldrich and used without further purification, and Toluene was dried over sodium. ^{1}H NMR and ^{13}C NMR spectra were recorded on Varian Oxford 300, 400 or 500 spectrometer at room temperature using tetramethylsilane as an internal standard. The reference values used for deuterated chloroform ($CDCl_3$) were 7.26 and 77.02 ppm for ^{1}H and ^{13}C NMR spectra, respectively. Chemical shifts (δ) are given in parts per million and coupling constants are given as absolute values expressed in Hertz. HRMS was obtained using a Micromass/Waters LCT instrument. Infra-red specra were recorded on a Varian 3100 FT-IR Excaliber Series spectrometer. Optical rotation values were measured on a Perkin Elmer 241 Polarimeter. $[\alpha]_D$ values are given in 10^{-1} deg cm^2 g^{-1}. HPLC analysis was carried out using a Supelco 2-4304 beta-Dex$^{®}$ 120 (30 m $\times$ 0.25 mm, 0.25 mm film) and a Chiralcel OD column (0.46 cm I.D. $\times$ 25 cm), respectively. Flash chromatography was carried out using Merck Kiesegel 60 F254 (230–400 mesh) silica gel. Evaporation in vacuo refers to the removal of volatiles on a Büchi rotary evaporator with an integrated vacuum pump. Thin-layer chromatography (TLC) was performed on Merck DC-Alufolien plates pre-coated with silica gel 60 F254. They were visualized either by quenching with ultraviolet fluorescence, or by charring with an acidic vanillin solution (vanillin, H_2SO_4 and acetic acid in MeOH). Preparative layer chromatography was carried out on glass plates pre-coated with silica gel HF$_{254+366}$ (Merck).

Contents

Symbols and Abbreviations

$[\alpha]_D^{20}$	specific rotation
AcOH	Acetic acid
AIBN	Azobisisobutyronitrile
app	apparent
aq	aqueous
Ar	aromatic
ATL	aspirin triggered Lipoxin
Bn	benzyl
br	broad
BuLi	butyl lithium
°C	degrees Celcius
COX	cyclooxygenase
δ	chemical shift in degrees downfield from TMS
d	doublet
DCM	dichloromethane
dd	double doublet
ddd	double double doublet
de	diastereomeric excess
DIP Cl	Chlorodiisopinocampheylborane
DMA	*N,N*-dimethylacetamide
DMAP	Dimethylaminopyridine
DMF	*N,N*-dimethylformamide
DMP	2,2-Dimethoxypropane
DMSO	dimethylsulfoxide
dppf	1,1′-Bis(diphenylphosphino)ferrocene
ee	enantiomeric excess
EOR	eicosanoid oxido-reductase
eq	equivalent
ESMS	electrospray mass spectrometry
Et	ethyl
EtOAc	ethyl acetate

EtOH	ethanol
g	gram (s)
gen	generation
GPCR	G-protein coupled receptor
h	hour (s)
HPLC	high pressure liquid chromatography
HRMS	high resolution mass spectroscopy
Hz	Hertz
IL	Interleukin
IFN	Interferon
IR	infrared spectroscopy
i-Pr	*iso*-propyl
J	coupling constant
LDA	lithium diisopropylamide
LO	Lipoxygenase
LTB_4DH	Leukotriene B_4 12-hydroxy dehydrogenase
LX	Lipoxins
LXA_4	Lipoxin A_4
LXB_4	Lipoxin B_4
m	multiplet
M	molar
M^+	molecular ion
MCP	monocyte chemoattractant protein
Me	methyl
MIP	Macrophage inflammatory protein
MeCN	acetonitrile
MeOH	methanol
mg	milligram
$MgSO_4$	magnesium sulphate
min	minute (s)
mL, μL	millilitre, microlitre
mol, mmol, μmol	mole, millimole, micromole
m.p.	melting point
NBS	*N*-bromosuccinimide
nM, μM	nano molar, micro molar
NaOAc	sodium acetate
NaOMe	sodium methoxide
OAc	acetate
o, m, p	*ortho, meta, para*
PCC	pyridinium chlorochromate
PG	prostaglandin
PGDH	prostaglandin dehydrogenase
PGR	prostaglandin reductase
Ph	phenyl
PMA	Phorbol 12-myristate 13-acetate

PMN	polymorphonuclear leukocytes
ppm	parts per million
PMP	1,2,2,6,6-pentamethylpiperidine
PPTS	pyridinium *p*-toluenesulfonate
p-TSA	*para*-Toluenesulfonic acid
RANTES	Chemokine (C-C motif) ligand 5
R_f	retention factor
r.t.	room temperature
s	singlet
SDF-1	stromal cell-derived factor-1
SEM	Standard error of the mean
t	triplet
TBAF	tetra-*n*-butylammonium fluoride
t-Bu	*tert*-butyl
TBDMS	*tert*-Butyldimethylsilyl
TEA	triethylamine
THF	tetrahydrofuran
THP-1	Human acute monocytic leukemia cell line
TLC	thin-layer chromatography
TMS	tetramethylsilane
TMSCl	trimethylsilyl chloride
TMSBr	trimethylsilyl bromide
TNF	tumor necrosis factor
^{1}H, ^{13}C NMR	^{1}H, ^{13}C nuclear magnetic resonance

Chapter 1
Introduction

Mankind have always used natural resources in an effort to treat a variety of human ailments. These resources, including natural products, have traditionally been sourced from plants, animals or microorganisms [1]. Nature has the ability to biosynthesise both simple and complex molecules which often have therapeutic effects [2]. Such natural products are commonly secondary metabolites, compounds synthesised from primary metabolites such as amino acids, after an often complex series of metabolic steps. Despite the elaborate metabolic pathways undertaken to produce these metabolites, the precise biological function of these compounds often remains a mystery. However, in many cases these entities are known to have beneficial effects for the host organism.

1.1 Secondary Metabolites in Medicine

Increased knowledge concerning the function of these secondary metabolites has led to their extensive exploration in modern medicine. Important therapeutic compounds have been inspired from metabolites isolated from natural sources such as the Willow bark and the Pacific Yew tree. These revolutionary drugs contain varying degrees of molecular architecture, from the widely used analgesic Aspirin, to the extremely successful anticancer agent Taxol, Fig. 1.1 [3, 4].

The isolation and structural characterisation of natural secondary metabolites is a challenging area of research. It has the ability to provide excellent lead drug candidates with 40% of modern drugs being developed from natural products [5]. Since nature can only supply these important compounds in minimal quantities, efficient synthetic routes are necessary for their preparation in order to avoid excess exploitation of natural resources and to satisfy commercial demand.

C. Duffy, *Heteroaromatic Lipoxin A₄ Analogues*, Springer Theses,
DOI: 10.1007/978-3-642-24632-6_1, © Springer-Verlag Berlin Heidelberg 2012

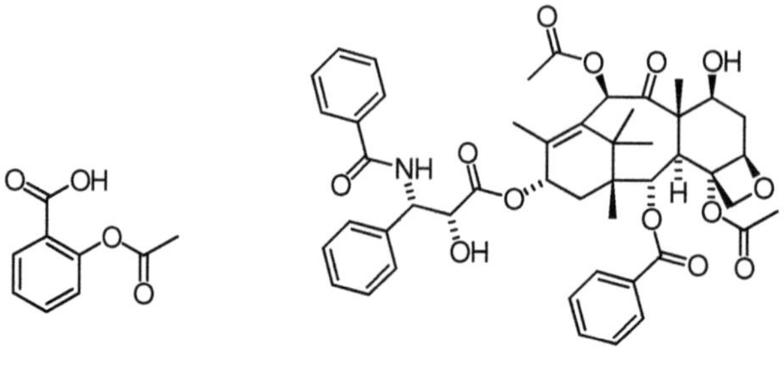

Aspirin   Taxol

Fig. 1.1 Therapeutic agents inspired by nature

1.2 Natural Products as Lead Compounds for Drug Discovery

Another excellent example of commercial drugs developed from natural products includes a class of compounds known as the statins. These molecules represent a remarkable class of cholesterol lowering agents which act as enzyme inhibitors [1]. The statins have been exploited extensively by the pharmaceutical industry as they reduce the risk of heart attacks and strokes [6, 7]. Mevastatin, Fig. 1.2, was the first molecule to be isolated from this class of compound and it became known as a Type I statin. Its isolation from *Penicillium citrinum* in 1970 sparked widespread interest as it was found to be an effective and potent statin [1].

However, largely due to problems in preclinical trials, Mevastatin never reached market. Nevertheless, it can be regarded as a landmark for drug discovery as it has paved the way for synthetic Type II statins.

Atorvastatin, Fig. 1.3, a Type II synthetic statin, is currently marketed as Lipitor and holds the position as the best selling drug worldwide. Presently, Lipitor sales are worth more than \$1 billion every month [8], demonstrating the significance of natural products as lead compounds for drug discovery.

1.3 Secondary Metabolites from Humans

An alternative strategy to drug design and discovery stems from investigating the secondary metabolites produced by humans instead of plants, animals or microorganisms. This rationale is inspired from the beneficial effects that the plant and microbial metabolites have on the host organism. Prostaglandins are a group of lipid mediators derived from the oxidation of C_{20} essential fatty acids [9]. They are produced on demand within the cell from arachidonic acid. These short lived messenger molecules have the ability to carry out numerous biological functions varying from inducing labour during childbirth to triggering pain and inflammation [10–12]. The prostaglandins are short lived as they are enzymatically converted to

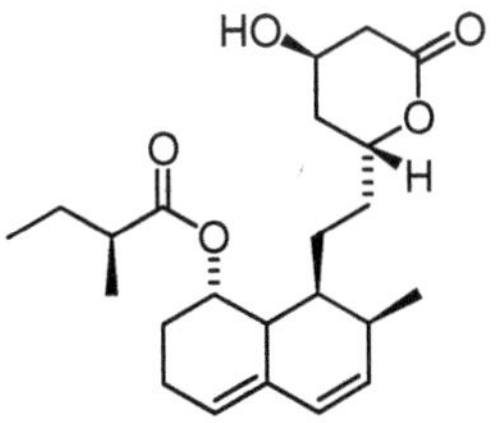

Isolated in 1970 from
Penicillium citrinum

Mevastatin

Type I

Fig. 1.2 Structure of type I statin, Mevastatin

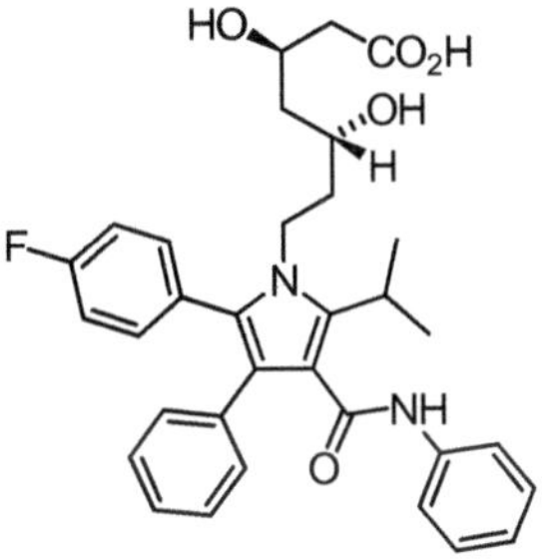

Lipitor - Best Selling
Drug Worldwide

Atorvastatin

Type II

Fig. 1.3 Structure of type II statin, Atorvastatin

inactive metabolites. Inspite of this, encouraging therapeutic agents originating from this class of compound have been exploited by the pharmaceutical industry. Xalatan, for example, is an effective drug used to treat ocular hypertension or glaucoma [11]. This drug, with an additional phenyl substituent on the lower chain, is far more effective than its $PGF_{2\alpha}$ derivative, as it has reduced ocular side effects such as irritation and conjunctival hyperemia, Fig. 1.4 [11].

1.4 Discovery and Isolation of Lipoxins

Another important class of secondary metabolites, oxygenated derivatives of arachidonic acid, were discovered and identified from human leukocytes by Serhan and Samuelsson in 1984 [13, 14]. Lipoxin A_4 (LXA$_4$) and Lipoxin B_4 (LXB$_4$), Fig. 1.5, are trihydroxytetraene-containing eicosanoids.

They are produced by the sequential actions of lipoxygenases (LO) during a series of complex cellular interactions [13]. LO are a family of iron-containing enzymes, which are known to catalyse the oxygenation of unsaturated fatty acids

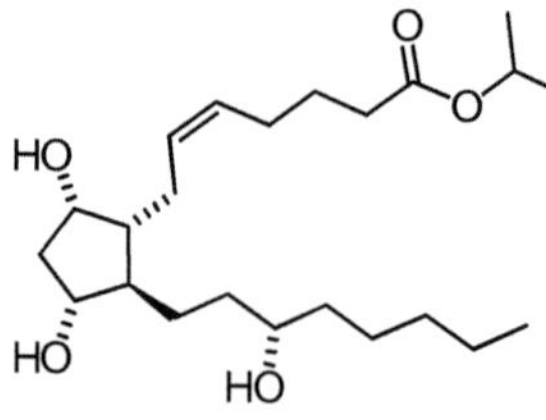

PGF$_{2\alpha}$ derivative 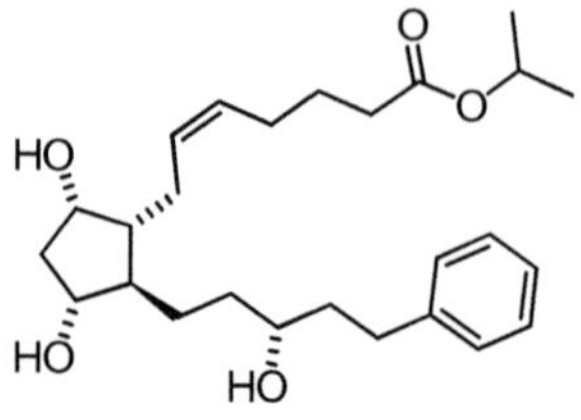   Xalatan

Fig. 1.4 Xalatan used to treat ocular hypertension or glaucoma

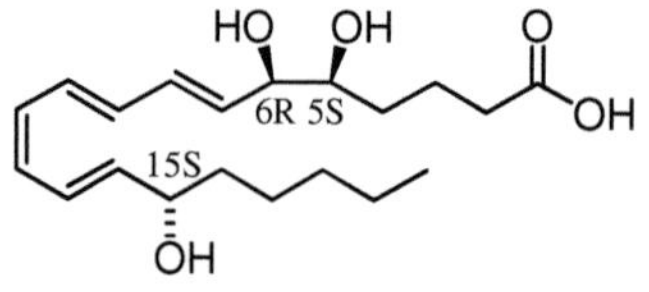

Lipoxin A$_4$ 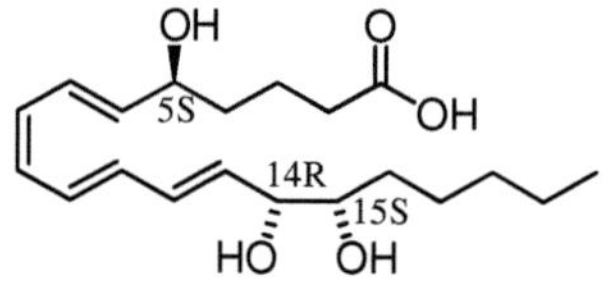   Lipoxin B$_4$

Fig. 1.5 Structures of Lipoxin A$_4$ (LXA$_4$) and Lipoxin B$_4$ (LXB$_4$)

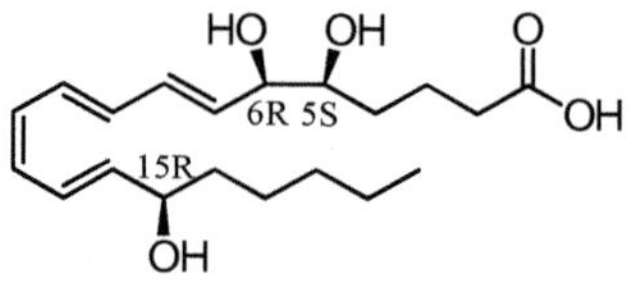

Aspirin triggered Lipoxin A$_4$

Fig. 1.6 Aspirin triggered LXA$_4$

and lipids. The combined oxygenase activity of 5-, 12- and 15-LO leads to the biosynthesis of LXA$_4$ and LXB$_4$ [15]. The 15-epi-LX or aspirin triggered Lipoxins (ATL), Fig. 1.6, differ only in the stereochemistry at C$_{15}$ and are produced by aspirin-acetylated cyclooxygenase-2 (COX-2) [16].

Once biosynthesised, enzymatically derived LXA$_4$ and LXB$_4$ are known to possess potent and selective anti-inflammatory activity [17]. They act as so called "stop signals" by activating the receptor ALXR to prevent the migration of neutrophils to sites of inflammation [18, 19]. During injury, an inflammatory response is triggered and a cascade of cellular events occur at the site of inflammation, which include the migration of neutrophils. These neutrophils accumulate usually within one hour of the injury and this event is regarded as the most important process in the lead up to inflammation [10].

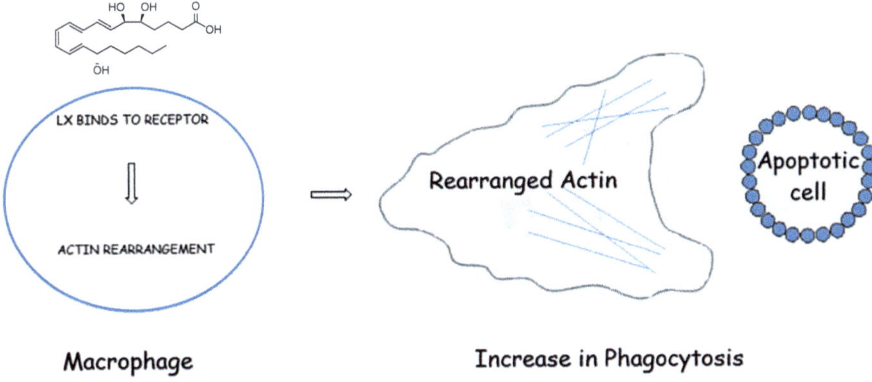

Fig. 1.7 Lipoxin mechanism of action [23]

Monocytes also accumulate at the site and develop into larger macrophages which cause the phagocytosis of apoptotic PMN. Lipoxins have previously been shown to possess the ability to regulate polymorphonuclear leukocytes (PMNs), chemotaxis, adhesion and transmigration [20]. It has been demonstrated that Lipoxins resolve inflammation by promoting nonphlogistic phagocytosis of apoptotic PMN by macrophages in vitro and in vivo [21]. Lipoxins bind via a specific G protein-coupled receptor, named ALX [18, 22]. G-Protein Coupled Receptors (GPCRs), also known as seven transmembrane receptors, are a large protein family of transmembrane proteins. They have the ability to sense molecules outside the cell and activate signal transduction pathways and ultimately, cellular responses. In this case, Lipoxins act as agonists by binding to the GPCR embedded in the cell membrane, which induces phagocytosis of neutrophils, thereby resolving inflammation, Fig. 1.7 [23].

1.5 Rapid Metabolism of Lipoxins

As previously mentioned, the accumulation of LXA_4 and LXB_4 at the site of inflammation is short lived. As with all autocoids, LX are rapidly metabolised in vivo into inactive metabolites, Scheme 1.1 [24].

Lipoxin A_4 is converted by specific leukocytes into 15-oxo-LXA_4, 13, 14-dihydro-15-oxo-LXA_4 and 13, 14-dihydro-LXA_4 and oxidation can also occur at C_{20}. This instability issue is a major obstacle to the application of these compounds as potential pharmacological agents.

The therapeutic potential of natural products, such as those outlined above, is hampered by issues such as instability and/or limited quantities of these natural resources. The pharmaceutical industry and academia have tried to overcome these hurdles in a number of ways. Firstly, they have developed efficient synthetic routes

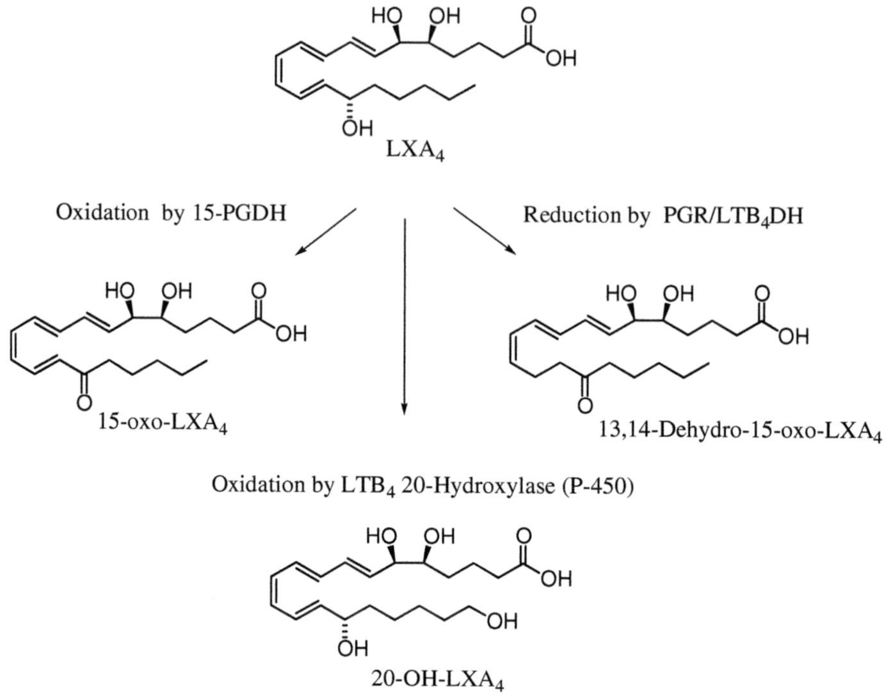

Scheme 1.1 Rapid metabolism of LXA$_4$ [24]

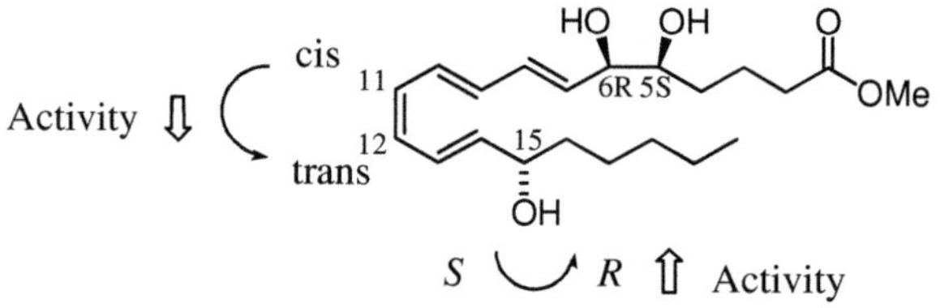

Fig. 1.8 Structure activity relationships

which allow them to prepare large quantities of the active natural ingredients. Secondly, they have designed analogues of these active natural products in order to enhance their bioactivity and improve their stability.

1.6 Structure Activity Relationships of Natural Lipoxins

Structure activity relationships of natural Lipoxins have been extensively reported which show certain functionalities and stereocentres are extremely important in order to retain biological activity, Fig. 1.8 [24, 25].

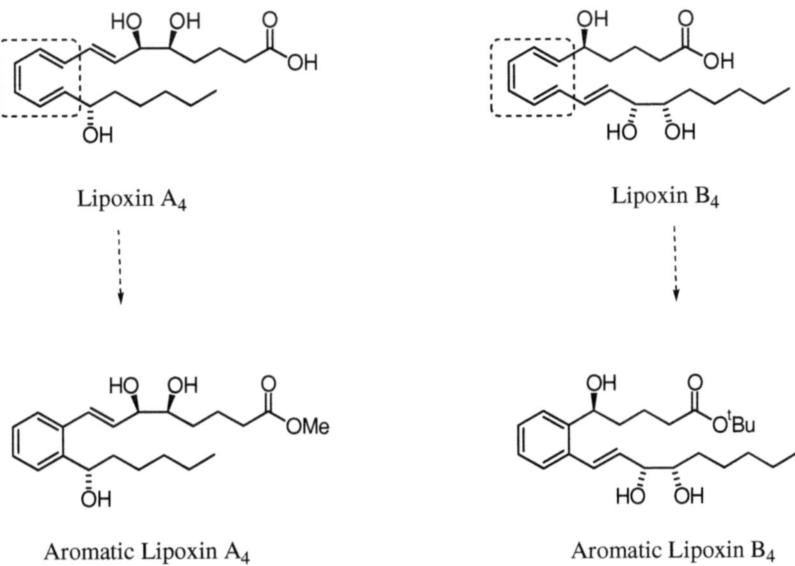

Lipoxin A$_4$

Lipoxin B$_4$

Aromatic Lipoxin A$_4$

Aromatic Lipoxin B$_4$

Fig. 1.9 Aromatic replacements of the native Lipoxin A$_4$ and Lipoxin B$_4$ [28]

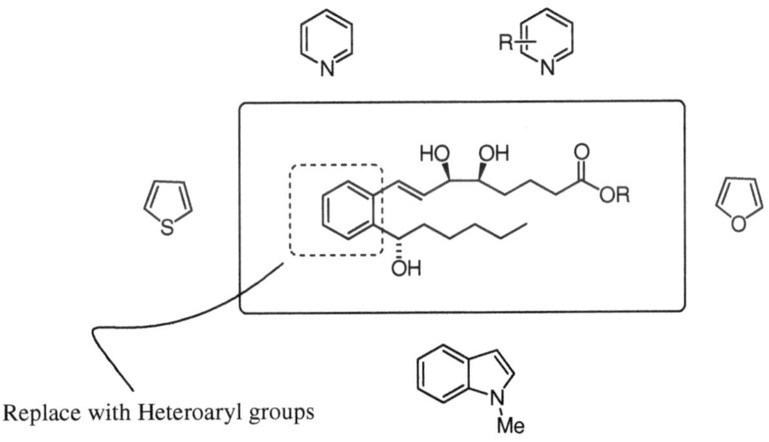

Fig. 1.10 Synthetic research objectives

Alcohols (5S and 6R) were found to be essential for retention of the bioactivity. Inversion of the chirality from S to R at C_{15} caused an increase in activity. Double bond isomerisation at C_{11-12} resulted in a significant decrease in activity in the biological potency.

Their isolation in 1984 prompted the search for new pharmacological drug candidates based on these potential therapeutic agents. As with many natural products, minimal quantities result from isolating these compounds from natural sources. This inspired the development of efficient synthetic routes for their preparation. Extensive spectroscopic and chromatographical evidence, combined with comparisons of biological activities, proved LXA$_4$ to be (5S,6R,15S)-trihyroxy-(7E,9E,11Z,13E)-icosatetraenoic acid [26], and LXB$_4$ to be (5S,14R,15S)-trihyroxy-(6E,8Z,10E,12E)-icosatetraenoic acid [27].

1.7 Design of Stable Lipoxin Analogues

Recently, our research group have focused their attention on modifying the triene structure of the Lipoxin A$_4$ and Lipoxin B$_4$ framework. We have successfully replaced the triene moiety of the native Lipoxin A$_4$ and Lipoxin B$_4$, Fig. 1.9 [28].

Replacement of this part of the molecule with a benzene ring has major advantages in terms of (i) considerably increasing the stability of the molecule towards enzymatic decomposition, (ii) development of a short and economical synthesis in an effort to access and screen numerous analogues to further tune the pharmacological profile, and (iii) prevention of the double bond isomerisation as described above. The synthesis and biological evaluation of these aromatic analogues will be discussed in more detail in Chap. 2.

1.8 Design of Heteroaromatic Lipoxin A$_4$ Analogues

The exciting results observed upon replacement of the enzymatically unstable native triene structure with benzene have stimulated an investigation into the synthesis of heteroaromatic LXA$_4$ analogues for biological evaluation. Replacement of the benzene ring with an array of heteroaryl groups has the potential to increase the pharmacological profile of these eicosanoids, Fig. 1.10.

In this context, we have designed a synthesis which allows us to replace the stable benzene moiety with five- and six-membered heterocycles such as thiophene and pyridine, respectively. Finally we attempted the synthesis of furan- and indole-containing LXA$_4$ analogues. These analogues will assist in our ongoing efforts to prepare and evaluate novel bioactive LXA$_4$ analogues with anti-inflammatory effects.

References

1. Patrick GL (2009) An introduction to medicinal chemistry, 4th edn. Oxford University Press, New York
2. Nicolaou KC, Vourloumis D, Winssinger N, Baran PS (2000) Angew Chem Int Ed Engl 39:44

3. Nicolaou KC, Yang Z, Liu JJ, Ueno H, Nantermet PG, Guy RK, Claiborne CF, Renaud J, Couladouros EA, Paulvannan K, Sorensen EJ (1994) Nature 367:630
4. Wothers P, Greeves N, Warren S, Clayden J (2001) Organic chemistry. Oxford University Press, New York
5. Sarker SD, Latif Z, Gray AI (2006) Natural products isolation, 2nd edn. Humana Press, Totowa
6. Crouse JR III, Byington RP, Furberg CD (1998) Atherosclerosis 138:11
7. Crouse JR, Byington RP, Bond MG, Espefand MA, Craven JW, Sprinkle JW, McGovern MF, Furberg CD (1995) Am J Cardiol 75:455
8. Wenner Moyer M (2010) Nat Med 16:150
9. Bergstrom S (1967) Science 157:382
10. Campbell NA, Reece JB (2001) Biology, 6th edn. Pearson Education, San Francisco
11. Collins PW, Djuric SW (1993) Chem Rev 93:1533
12. Corey EJ (1991) Angew Chem Int Ed Engl 30:455
13. Serhan C, Hamberg M, Samuelsson B (1984) Biochem Biophys Res Comm 118:943
14. Serhan C, Hamberg M, Samuelsson B (1984) Proc Natl Acad Sci U S A 81:5335
15. Serhan CN (2005) Prostaglandins Leukot Essent Fat Acids 73:141
16. Claria J, Serhan CN (1995) Proc Natl Acad Sci U S A 92:9475
17. McMahon B, Godson C (2004) J Renal Physiol 286:F189
18. Brink C, Dahlén S-E, Drazen J, Evans JF, Hay DWP, Nicosia S, Serhan CN, Shimizu T, Yokomiao T (2003) Pharmacol Rev 55:195
19. Chiang N, Serhan C, Dahlén S-E, Drazen JM, Hay DWP, Rovati GN, Shimizu T, Yokomiao T, Brink C (2006) Pharmacol Rev 58:463
20. Godson C, Mitchell S, Harvey K, Petasis NA, Hogg N, Brady HR (2000) J Immunol 164:1663
21. Mitchell S, Thomas G, Harvey K, Cottell DC, Reville K, Berlasconi G, Petasis NA, Erwig L, Rees AJ, Savill J, Brady HR, Godson C (2002) J Am Soc Nephrol 13:2497
22. Maddox JF, Hachicha M, Takano T, Petasis NA, Fokin VV, Serhan CN (1997) J Biol Chem 272:6972
23. Scannell M, Maderna P (2006) Sci World 6:1555
24. Clish B, Levy D, Chiang N (2000) J Biol Chem 275:25372
25. Chiang N, Fierro I, Gronert K, Serhan C (2000) J Exp Med 191:1197
26. Serhan CN, Nicolaou KC, Webber SE, Veale CA, Dahlén S-E, Puustinen TJ, Samuelsson B (1986) J Biol Chem 261:16340
27. Serhan CN, Hamberg M, Samuelsson B, Morris J, Wishka DG (1986) Proc Natl Acad Sci U S A 83:1983
28. O' Sullivan TP, Vallin KSA, Shah STA, Fakhry J, Maderna P, Scannell M, Sampaio ALF, Perretti M, Godson C, Guiry PJ (2007) J Med Chem 50:5894

Chapter 2
Recent Advances in the Chemistry and Biology of Stable Synthetic Lipoxin Analogues

2.1 Introduction

The Lipoxin metabolites, discussed in Chap. 1, dramatically reduce the bioactivity of this class of compounds and render them poor potential pharmacological agents. In light of the findings associated with the stabilisation and market value of the synthetic prostaglandin and prostacyclin analogues [1], it was thought that a similar approach could be beneficial with respect to the native Lipoxins (LX).

2.2 Design, Synthesis and Biological Evaluation of Stable Lipoxin Analogues

Recent synthetic efforts include mimicking the core structure of the native LXA_4 **1** by replacing certain functionalities with chemically stable motifs with the aim of retaining the potent biological activity. These stable analogues will be sub-divided into three distinct categories (**A**, **B** and **C**), based on the target area being modified, Fig. 2.1. The strategies include (**A**) structural modifications of the C_{15-20} chain: [2] (**B**) replacement of the triene with chemically stable aromatic/heteroaromatic systems: [3, 4] and (**C**) modifications of the C_{1-8} unit [5]. While excellent reviews have extensively covered the synthesis and biological relevance of the native LX and their stereoisomers [6, 7], this chapter will focus on the synthesis and biological evaluation of enzymatically durable analogues.

2.3 (A) Structural Modifications of the C_{15-20} Chain

The desire to prevent oxidation at C_{15-20} led to the design of the first LXA_4 analogues which showed resistance to oxidation [8]. Replacement of this alkyl chain with several different groups furnished a number of analogues with increased pharmacokinetic

C. Duffy, *Heteroaromatic Lipoxin A_4 Analogues*, Springer Theses,
DOI: 10.1007/978-3-642-24632-6_2, © Springer-Verlag Berlin Heidelberg 2012

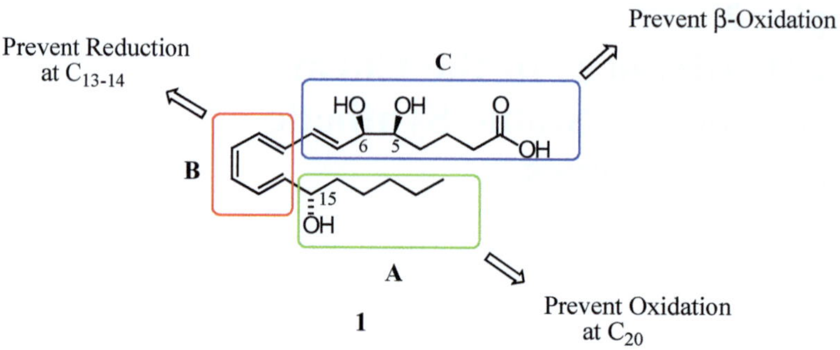

Fig. 2.1 Targeted domains for modifications of the native Lipoxin A_4 **1**

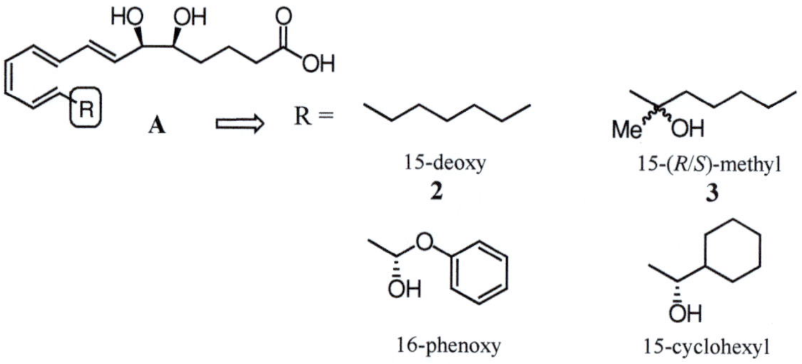

Fig. 2.2 Design of C_{15-20} stable analogues [8]

profiles. Structural adaptations incorporated 15-deoxy-LXA$_4$ **2**, 15-(*R/S*)-methyl **3**, 16-phenoxy **4**, and 15-cyclohexyl **5** into the C_{15-20} chain, Fig. 2.2.

The synthetic routes used for these analogues were not reported in the literature, although they were clearly constructed by using previously reported syntheses for the related native LX [9]. The authors observed that these structural modifications dramatically increased biostability compared to the native LX by preventing dehydrogenation by differential HL-cells and recombinant 15-hydroxyprosta-glandin dehydrogenase. The bioactivity was also secured in the 15-(*R/S*)-methyl **3**, 16-phenoxy **4** and 15-cyclohexyl **5** analogues due to their ability to prevent PMN transmigration and adhesion in leukocyte migration. The 15-deoxy-LXA$_4$ **2** showed the least activity suggesting that the hydroxyl group at C_{15} is essential for the preservation of bioactivity.

Alternative analogues have been developed which resulted in enhanced bio-activity compared to the native LX. These designs include the addition of a fluoro **6** and trifluoromethyl **7** group onto the 16-phenoxy analogue **4**, Fig. 2.3 [7, 10].

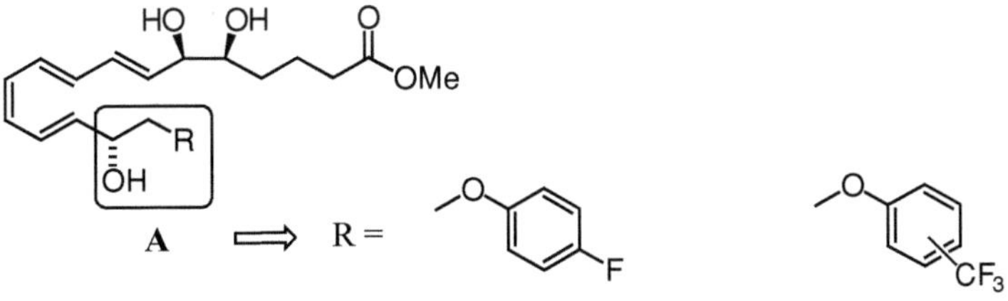

para-fluorophenoxy *ortho*-or *para*-trifluoromethylphenoxy

6 **7**

Fig. 2.3 Fluoro and trifluoromethyl stable analogues [7, 10]

The *para*-fluorophenoxy analogue **6** has proven itself to be an extremely potent derivative as it inhibited tumor necrosis factor (TNF)-α-induced leukocyte recruitment into the dorsal air pouch [10]. It was also found to suppress both LTB_4- and PMA-induced recruitment, when applied to mouse ear skin. Furthermore, this analogue has shown potential as an anti-cancer agent, as it inhibits endothelial cell proliferation leading to suppressed angiogenesis at the 1–10 nM range [11]. Realising the potential of these fluorinated analogues, a number of research groups began to develop efficient synthetic routes to these biologically important derivatives. The key synthetic transformations combine a *cis*-reduction of an alkyne, a palladium-catalysed Sonogashira reaction and a Wadsworth–Emmons alkene transformation, Scheme 2.1. Phillips and co-workers reported the first synthesis of the *para*-fluorophenoxy analogue **6** by adopting a chiral pool strategy [2], starting from 2-deoxy-D-ribose **11** [12]. This approach has the advantage of using a readily available starting material which incorporates the two stereocentres which will ultimately appear at C_5 and C_6.

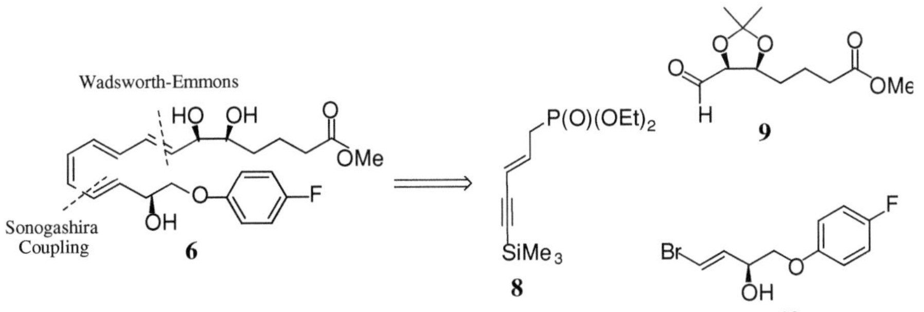

Scheme 2.1 Retrosynthetic analysis of *para*-fluorophenoxy analogue **6**

Protection of 2-deoxy-D-ribose **11** was achieved through its propylidine acetal **12** using 2-methoxypropene and pyridinium *p*-toluenesulfonate (PPTS) in ethyl acetate at room temperature, giving a 43% yield, Scheme 2.2. A Wittig reaction of methyl(triphenylphosphoranylidine)acetate and the aldehyde form of **12,** followed by a catalytic hydrogenation using 10% Pd/C furnished alcohol **13** in high yields of 81 and 87%, respectively. Oxidation of **13** using Swern conditions afforded aldehyde **9** in 86% yield. This was subjected to a Wadsworth–Emmons transformation with phosphonate **8** and deprotected using KF and 18-crown-6 to form the key intermediate **14** in 99% yield.

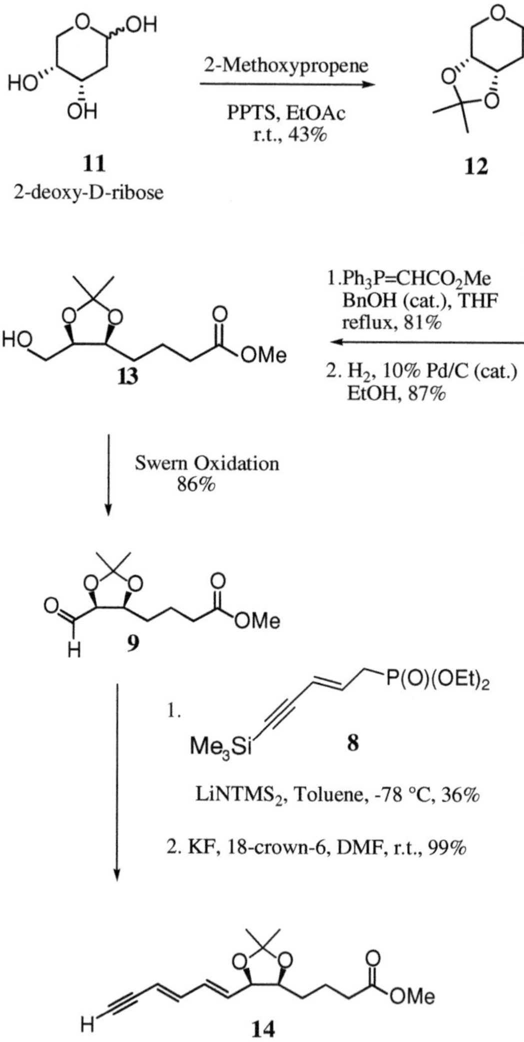

Scheme 2.2 Formation of key intermediate **14** [2]

Phosphonate **8** was itself assembled by the treatment of alkyne **15** with ethylmagnesium chloride and quenching with chlorotrimethylsilane followed by an Appel-type reaction gave the corresponding bromide in 90 and 74% yields, respectively, Scheme 2.3. This bromide was subjected to Arbusov reaction conditions to afford **8** in 90% yield.

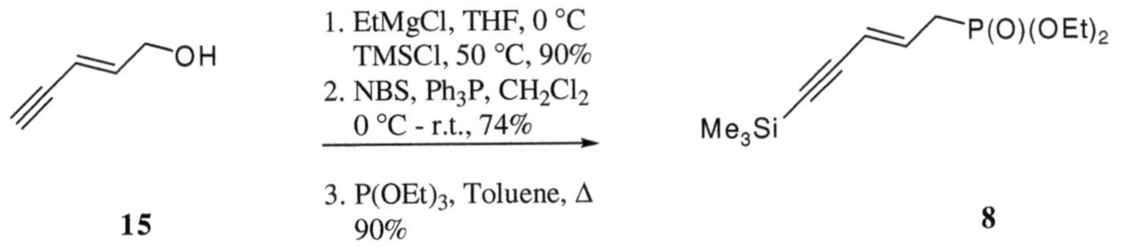

Scheme 2.3 Formation of phosphonate **8** [2]

The synthesis of the Sonogashira coupling partner **10** was accomplished in five steps, beginning with the alkylation of *p*-fluorophenol with 3-chloropropane-1,2-diol **16** in 56% yield, Scheme 2.4. Cleavage of the diol with silica-supported sodium periodate in dichloromethane afforded aldehyde **17** in 98% yield. Addition of lithium 2-trimethylsilylacetylide to **17**, followed by treatment with NaOH to

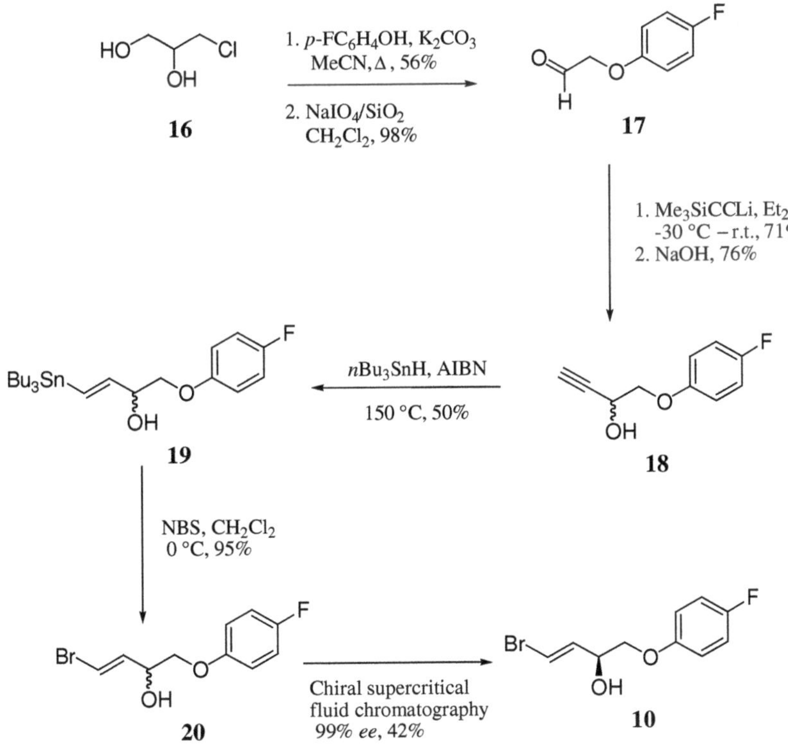

Scheme 2.4 Synthesis of Sonogashira coupling partner **10** [2]

remove the TMS group, gave alkyne **18** in 76% yield. Vinylstannane **19** was constructed by treating **18** with tri-*n*-butyltin hydride. Addition of NBS in dichloromethane to **19** gave the vinylbromide **20** in 95% yield. An attempted kinetic resolution of vinylstannane **19** using Sharpless epoxidation, followed by treatment of the unreacted alcohol with NBS to give **10**, proceeded with poor *ee*. Subsequently racemic **20** was resolved with chiral supercritical fluid chromatography to give vinylbromide **10** in 42% yield and 99% *ee*.

The Sonogashira reaction, employing $Pd(PPh_3)_4$ and CuI in the presence of *n*-propylamine at room temperature, was used to cross-couple vinylbromide **10** and the terminal alkyne **14**, resulting in the formation of **21** in 75% yield, Scheme 2.5. The catalyst loading was not given for this Sonogashira coupling. The acid sensitive acetal group was cleaved by the addition of methanolic HCl to give the corresponding diol. At this stage Lindlar's catalyst can be employed to access the C_{11-12} *cis*-double bond. However, problems have arisen with this method including over-reduction and isomerisation of the C_{11-12} *trans*-double bond isomer during the synthesis of other Lipoxin analogues [13]. Selective *cis*-reduction with an activated zinc alloy has previously been described by Boland [14], and this protocol afforded the *para*-fluorophenoxy Lipoxin analogue **6** in 80% yield. Activation of the zinc requires the addition of 2N HCl for 1–2 min for a clean reaction to take place.

In a similar synthetic approach, starting from 2-deoxy-D-ribose **11**, Petasis and co-workers synthesised stable Lipoxin analogues varying at the C_{15-20} chain, via the introduction of aliphatic, aromatic and fluoroaromatic groups, Scheme 2.6 [7]. The synthetic strategy incorporates a Wittig reaction for the construction of the C_{7-8} double bond, a Sonogashira reaction followed by a *cis*-reduction of the alkyne to establish the C_{11-12} double bond. Simple structural variations of the Sonogashira coupling partners gave rise to many synthetic analogues.

The precise details of the synthesis, including % yields and mol% of catalysts, were not reported as this was part of a review article. The *tert*-butyldimethylsilyl-protected aldehyde **22** was accessed through the chiral pool strategy using 2-deoxy-D-ribose **11**. Compound **23**, previously prepared [13], was reacted with **22** in a Wittig reaction. Double bond isomerisation with I_2 in dichloromethane followed by removal of the trimethylsilyl group by $AgNO_3$ and KCN in EtOH, THF and H_2O gave the alkyne coupling partner **24**. Reaction conditions employed for the Sonogashira reaction included $Pd(PPh_3)_4$, CuI in *n*-propylamine followed by the addition of the corresponding vinyl bromide or iodide. The *tert*-butyldimethylsilyl protecting groups were cleaved using TBAF in THF, followed by reduction of the alkyne, by either H_2 in the presence of Lindlar's catalyst, or by selective *cis*-reduction with an activated zinc alloy, to afford the series of analogues **26**. The 15-cyclohexyl, 15-cyclooctyl and the 16-phenoxy analogues were all found to retain the native Lipoxin bioactions. The inactivation by 15-PDGH and P-450-mediated ω-oxidation were hindered due to the absence of the free ω-alkyl chain. These analogues, of type **26**, were also found be extremely useful in studying the exact binding site in vivo [15]. The fluorinated analogues were found to be the most stable and active in vivo [10].

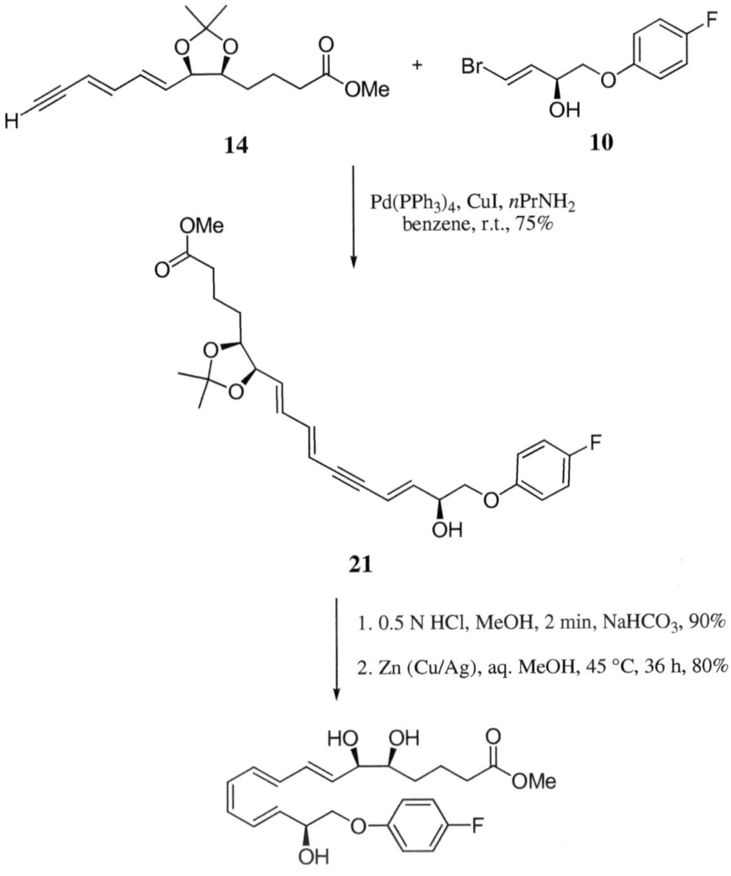

Scheme 2.5   Synthesis of *para*-fluorophenoxy Lipoxin analogue **6** [2]

2.4 (B) Structural Modifications of the Triene

In recent years, researchers have focused their attention on modifying the triene structure of the Lipoxin A_4 and B_4 framework. Derivitisation of this part of the molecule has major advantages in terms of (i) considerably increasing the stability of the molecule towards enzymatic decomposition (ii) development of a short and economical synthesis in an effort to access and screen numerous analogues to further tune the pharmacological profile and (iii) prevention of the double bond isomerisation as described above. Significant advances in the area include the substitution of the triene with aromatic [3, 4] and heteroaromatic rings [16], Fig. 2.4.

The LXA$_4$ and LXB$_4$ analogues reported by Guiry and co-workers, **27** and **28** respectively, were constructed using Sharpless asymmetric epoxidation,

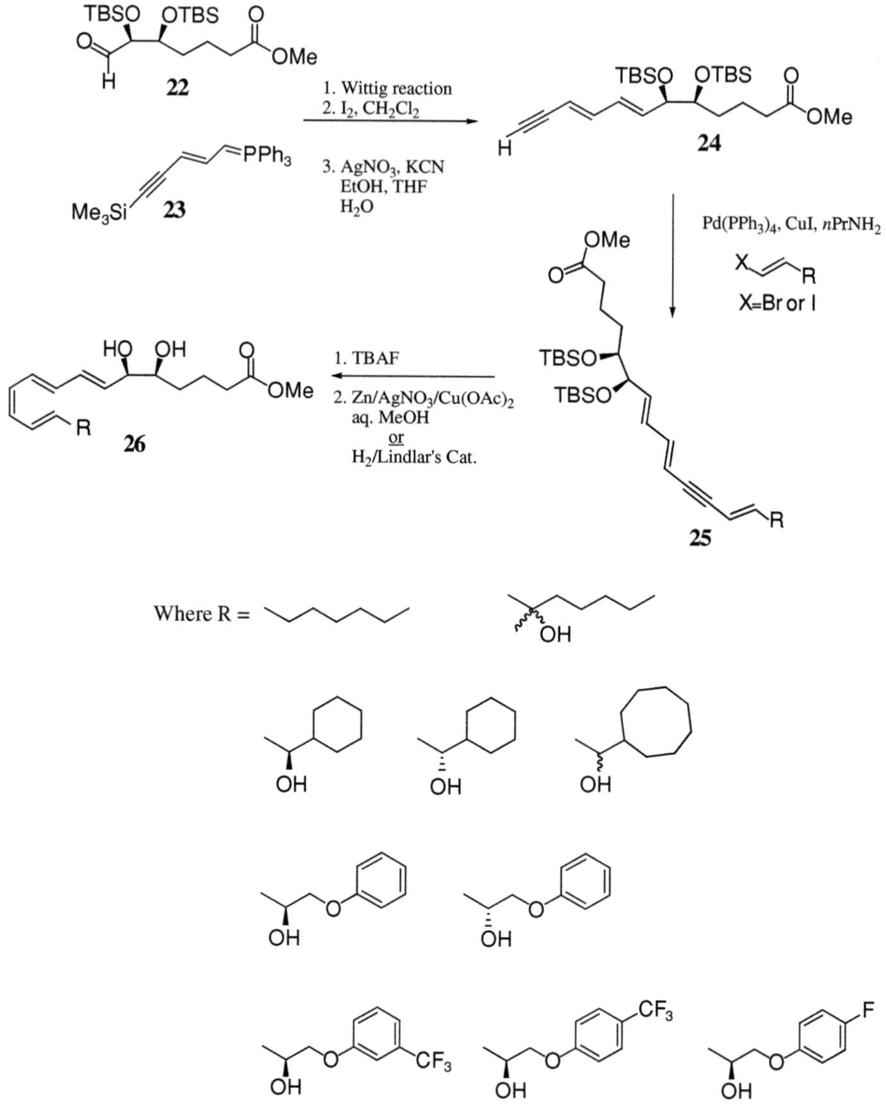

Scheme 2.6 Synthesis of aliphatic, aromatic and fluoroaromatic LXA$_4$ analogues [7]

Pd-mediated Heck coupling and diastereoselective reduction reactions as the key synthetic transformations [3]. These reactions provided enantio- and diastereoselective generation of each stereocentre and complete control for the formation of the *trans* olefin. In a similar synthetic route Guiry and co-authors synthesised a novel pyridine-containing LXA$_4$ **29** that was also found to possess important biological properties. The synthesis and biological evaluation of this pyridine-containing LXA$_4$ **29** will be discussed in detail in Chap. 4.

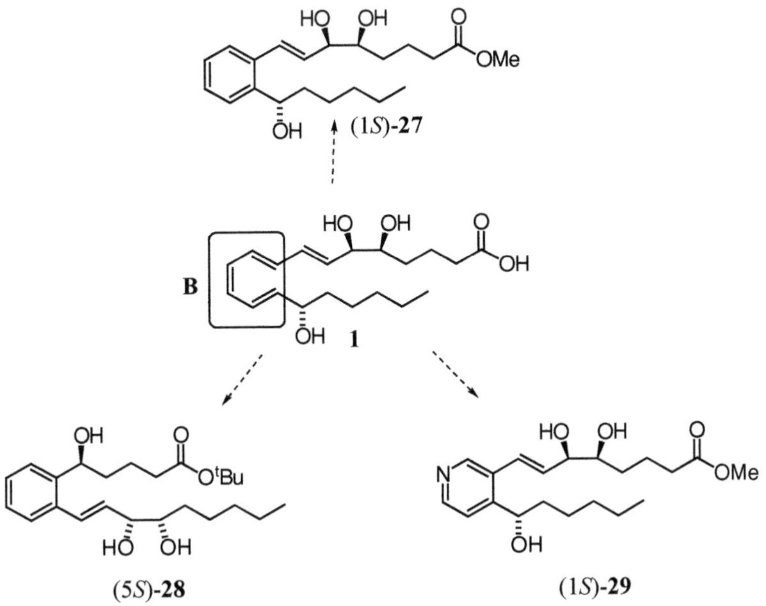

Fig. 2.4   Design of benzene- and pyridine-containing Lipoxin analogues [3]

The first stereoselective route to the novel aromatic analogue **27** described by Guiry and co-workers employed the commercially available divinylcarbinol **30** as the starting material, Scheme 2.7 [3].

This allylic alcohol **30** was subjected to Sharpless asymmetric epoxidation reaction conditions to give the chiral epoxide **31** in 85% yield and with an enantiomeric excess of greater than 99%. Ring opening of **31** with the Grignard derivative of **32** in the presence of a catalytic amount of CuI afforded the desired diol **33** in 82% isolated yield. This diol required an acid stable protecting group as the acidic Jones' reagent was applied to cleave the dioxane in the following transformation. The diol protection was successfully achieved by the addition of acetyl chloride and pyridine in THF at 0°C to give the bisacetate in 97% yield. The addition of Jones' reagent in acetone for 2 h yielded the corresponding acid **34**, which was esterified using diazomethane in diethyl ether. A change of protecting group strategy was employed at this stage as the *bis*-acetate methyl ester was an unsuitable coupling partner for the Heck reaction. For this reason, deprotection with NaOMe in MeOH followed by reprotection with a *tert*-butyldimethylsilyl group was necessary in order to afford the bis-silyl ether **35** in high yield. This protected olefin was then successfully applied in a palladium-mediated Heck reaction in both the benzene- and pyridine-containing LXA$_4$ analogues, **27** and **29**, respectively. The authors also found that zirconium tetrachloride was an efficient catalyst for a one-pot protection/deprotection synthetic methodology and used this for the synthesis of **35** [17]. This protocol also led to the synthesis of 6-acetoxy-5-hexadecanolide, a component of mosquito oviposition attractant pheromones [18], and also a microwave-assisted asymmetric synthesis of *exo*- and *endo*-brevicomin [19].

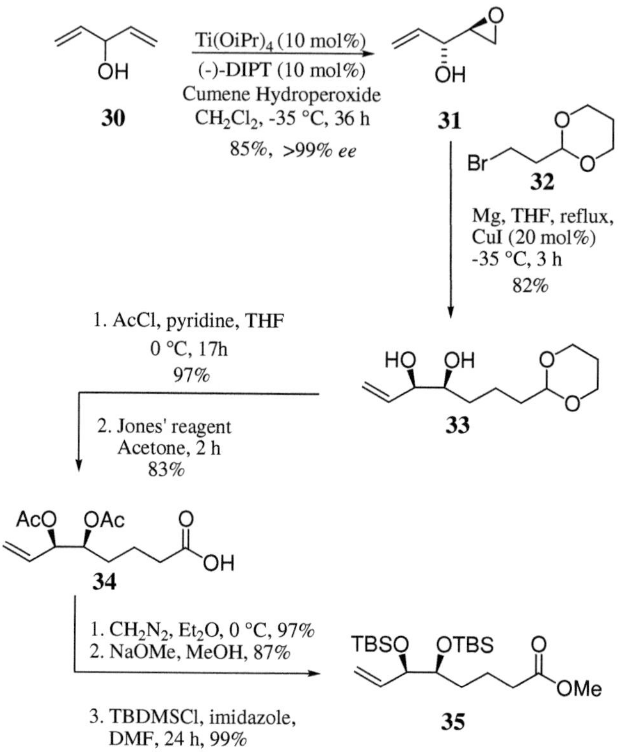

Scheme 2.7 Synthesis of key intermediate **35** [3, 16]

The preparation of aryl bromide **38** required as the other Heck coupling partner was achieved through the addition of the Grignard derivative of 1-brompentane **37** to acid chloride **36**, Scheme 2.8. The reaction was performed at −78°C to prevent any of the double addition product forming. An initial screening of Heck reaction conditions revealed that tributylamine, with its high boiling point, afforded the coupled product **39** in a very high yield (88%). Reduction of this ketone was achieved using sodium borohydride giving rise to a mixture of epimeric alcohols which were easily separated by column chromatography. The authors also employed (−)-β-chlorodiisopinocampheylborane to give alcohol **40** in 67% yield and with a 92% diastereomeric excess. Finally this alcohol was deprotected using *p*-toluenesulfonic acid in MeOH giving the triol (1*S*)-**27** in 84% yield. This triol and the (1*R*)-**27** analogue were both converted to their corresponding acids by LiOH in a mixture of methanol and water and were also investigated for their ability to aid in the resolution of inflammation.

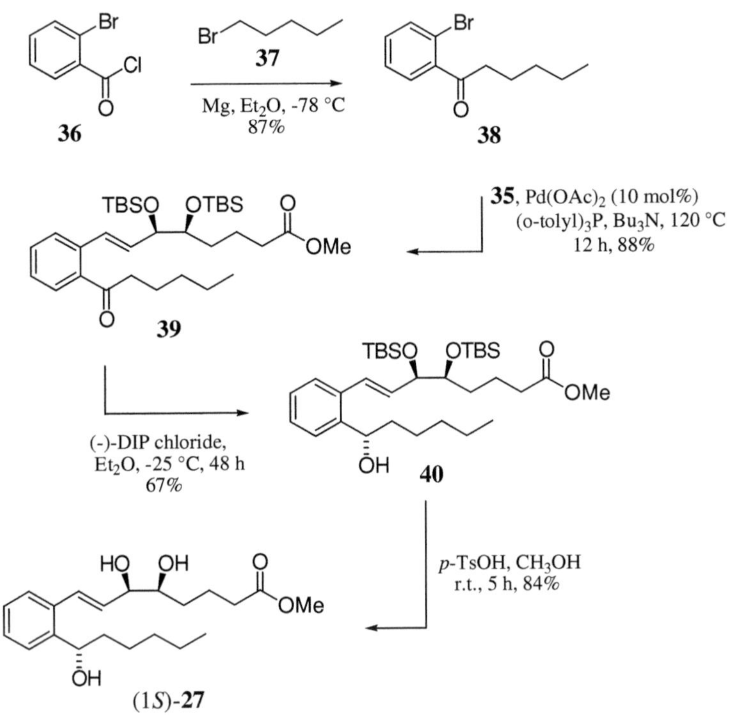

Scheme 2.8 Synthesis of aromatic LXA$_4$ (1*S*)-**27** [3]

The stereoselective synthesis of the aromatic LXB$_4$ analogue (5*S*)-**28** exploited a similar synthetic route, assembling the *trans* double via a palladium-catalysed Heck reaction with aryl bromide **43**, Scheme 2.9. The aryl bromide **43**, required for the Heck reaction, was formed through a Sonogashira coupling of 1-bromo-2-iodobenzene **41** and the commercially available terminal alkyne **42**, followed by oxidation with sulfonic acid and esterification.

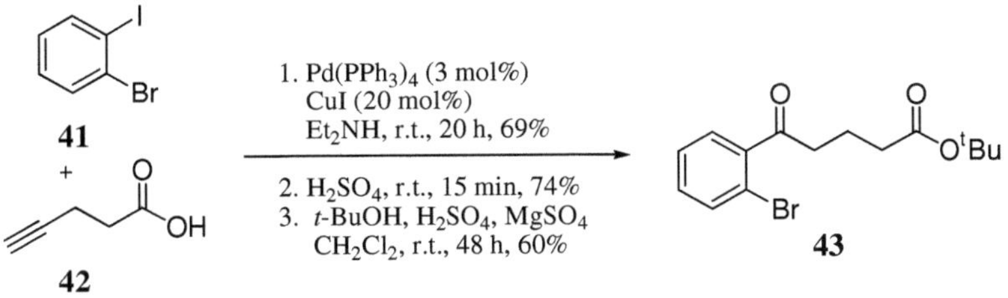

Scheme 2.9 Synthesis of Heck coupling partner **43** [3]

Another epoxide ring opening reaction via Grignard chemistry produced the olefin Heck coupling partner **44**, Scheme 2.10.

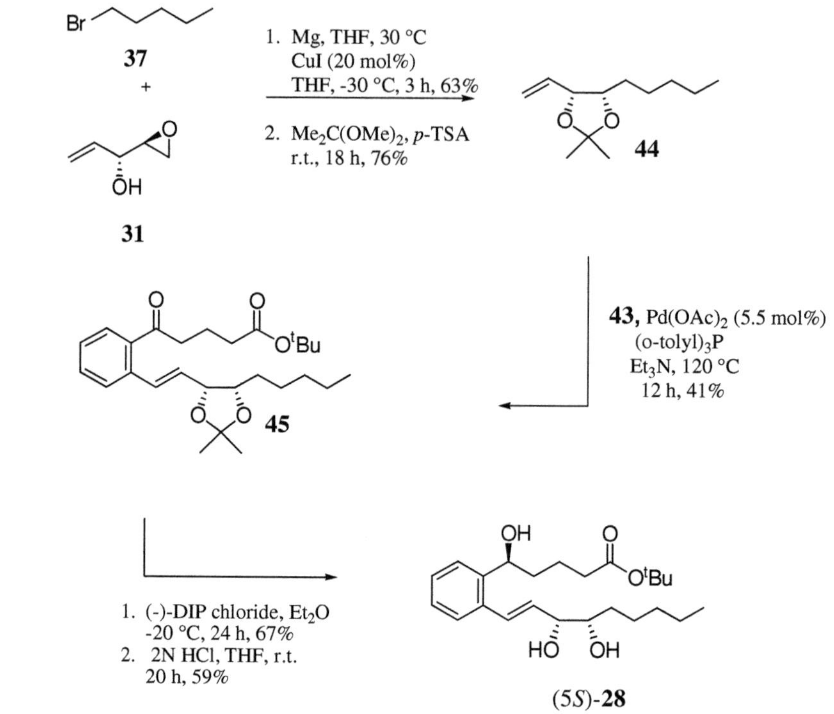

Scheme 2.10 Stereoselective synthesis of aromatic LXB$_4$ analogue (5*S*)-**28** [3]

The Heck reaction proceeded under similar reaction conditions to those employed for the synthesis of aromatic LXA$_4$ (1*S*)-**27**, furnishing **45** in 41% yield. Asymmetric reduction of ketone **45** was again accomplished by way of Brown's (–)-β-chlorodiisopinocampheylborane to give the alcohol in 67% yield with a de value of 97%. The final step was acetal cleavage using 2N HCl in THF at room temperature to furnish triol (5*S*)-**28** in 59% yield. These new aromatic analogues possess great potential as therapeutic agents as the modular synthetic approach to these compounds renders them extremely accessible and their pharmacodynamics can be further tuned by the addition of known classical bioisosteres.

The novel aromatic LXA$_4$ analogues (1*S*)-**27** and (1*R*)-**27** promoted increased clearance of apoptotic PMNs when compared to the effect of the native LXA$_4$, Fig. 2.5.

The aromatic LXB$_4$ (5*S*)-**28** analogue also stimulated phagocytosis of apoptotic PMNs with a maximum effect observed at 10^{-11} M, Fig. 2.6.

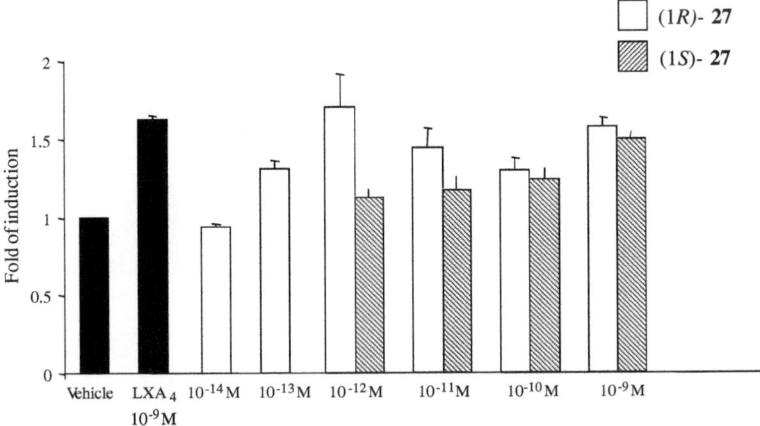

Fig. 2.5 Effect of (1*S*)-**27** and (1*R*)-**27** on the clearance of apoptotic PMNs [3]

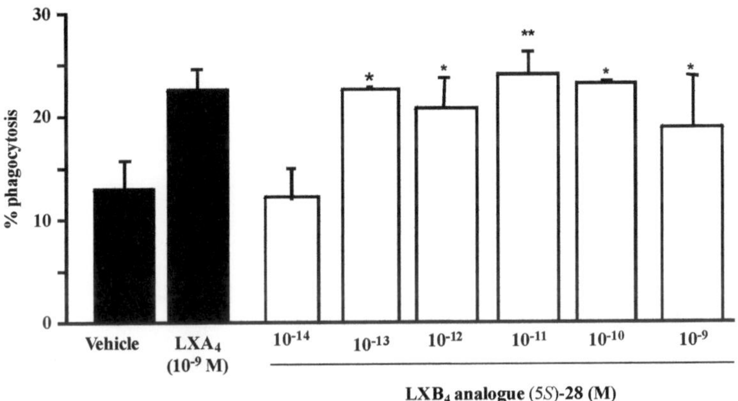

Fig. 2.6 Effect of (5*S*)-**28** on the clearance of apoptotic PMNs [3]

In addition to this, both analogues (**27** and **28**) caused F-actin rearrangement which has also been observed with the native compounds, Fig. 2.7 [20].

Phagocytosis of PMNs was inhibited by pre-treatment with the pan-FPR inhibitor Boc2. This strongly suggests that the effect of these analogues is mediated by the activation of the LX receptor, Fig. 2.8.

These analogues were also screened for their ability to stimulate adherence of monocytes to a matrix such as laminin, Fig. 2.9, which is a previously known property of the native LX and also some of the synthetically stable analogues [21, 22]. In the experiments the acids did not exhibit an increase in phagocytosis over the same concentration range as the methyl esters [3]. This lack of activity

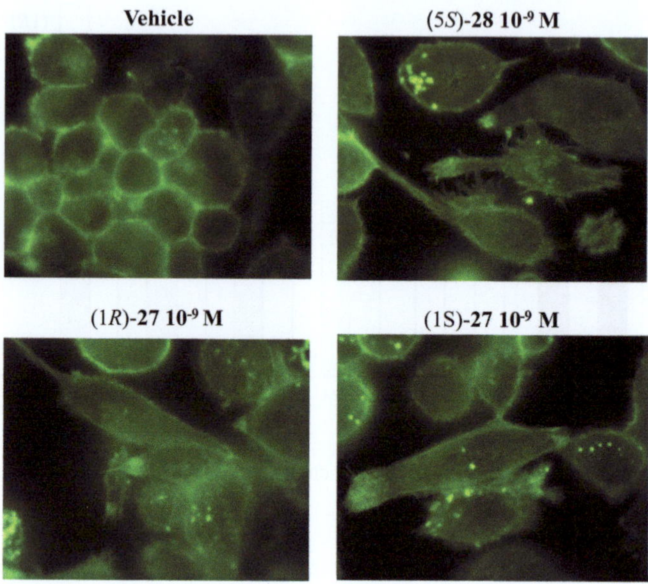

Fig. 2.7   Effect of LX analogues on actin rearrangement inTHP-1 cells [3]

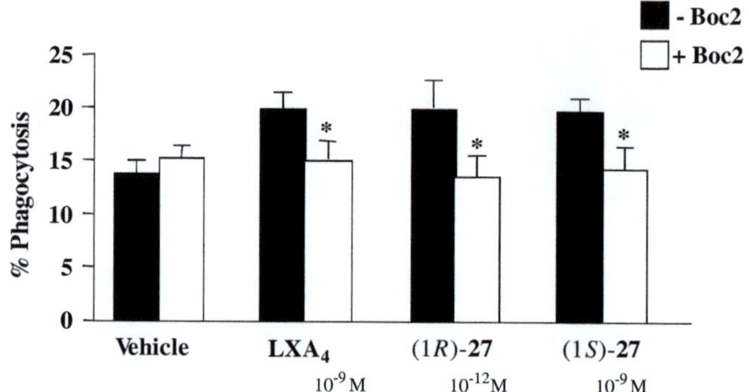

Fig. 2.8   LXA$_4$ analogues-stimulated phagocytosis of apoptocic PMNs is blocked by the rececptor antagonist BOC2 [3]

was attributed to the fact that the esters act as prodrugs, converting in vivo to the free acid and evoking LX-mediated biological actions [23].

Bannenberg and co-workers also showed that oral administration of LXA$_4$ has the ability to inhibit leukocyte infiltration in zymosan A-induced peritonitis [24]. Guiry and co-workers found that their (1R)-**27** analogue caused a significant decrease in neutrophil accumulation at 50 µg/kg while the (1S)-**27** analogue also showed a decrease at the highest dose tested, Fig. 2.10.

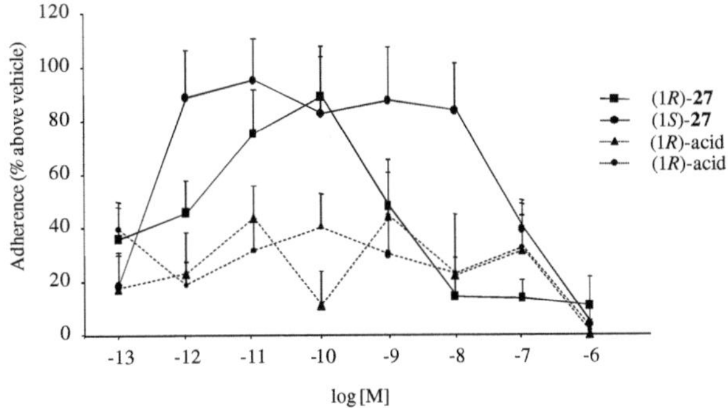

Fig. 2.9 Effect of stable analogues on THP-1 cell adherence to laminin [3]

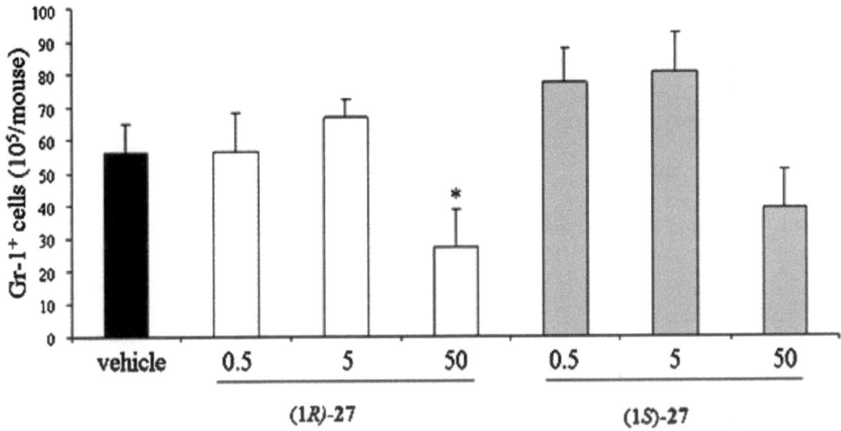

Fig. 2.10 Effect of LXA$_4$ analogues on zymosan-induced peritonitis [3]

Petasis and colleagues have also successfully managed to stabilise the native LXA$_4$ **1** with the same approach, replacement of the triene with a more durable benzene ring [4, 25]. Their synthetic route allowed for the synthesis of an array of analogues (**27, 46–49**) Fig. 2.11. Compounds **46–49** were designed from a strategy combining domain modifications (**A**) and (**B**), Fig. 2.1.

The synthesis of **46** and **47** relied on two sequential Suzuki–Miyaura coupling reactions, Scheme 2.11. The first combines 2-bromophenylboronic acid **51** and vinyl iodide **50**, which was constructed by a Takai olefination of **22** [13]. Suzuki–Miyaura reaction conditions incorporated Pd(PPh$_3$)$_4$ and K$_2$CO$_3$ using dioxane as the solvent at 60°C furnished **52** in 70% yield. The catalyst loading was not reported in this coupling reaction.

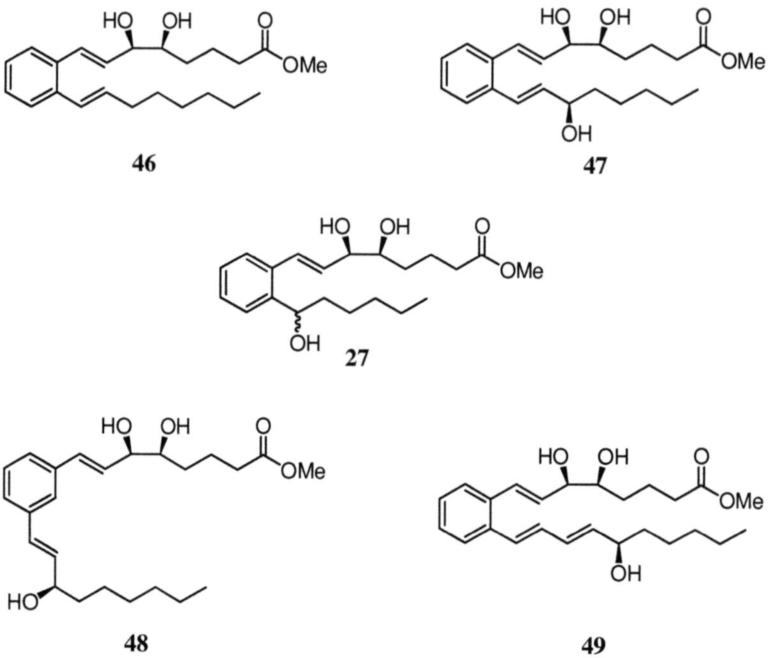

46 **47**

27

48 **49**

Fig. 2.11 Analogues designed and synthesised by Petasis and co-workers [4]

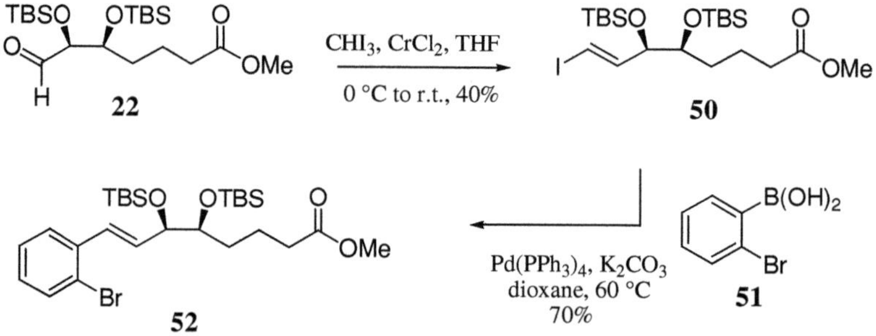

CHI$_3$, CrCl$_2$, THF
0 °C to r.t., 40%

22 **50**

Pd(PPh$_3$)$_4$, K$_2$CO$_3$
dioxane, 60 °C
70%

52 **51**

Scheme 2.11 Synthesis of key intermediate **52**

Boronic esters **55** and **56** were both synthesised from the corresponding alkynes **53** and **54**, respectively, Scheme 2.12. Compound **54** was synthesised by the protection of the corresponding alcohol [26, 27].

The second Suzuki–Miyaura coupling combined aryl bromide **52** and boronic esters **55** and **56** in the presence of Pd(PPh$_3$)$_4$ and K$_2$CO$_3$ using a mixure of dioxane and water as the solvent at 80°C, giving **57** and **58** in moderate yields, Scheme 2.13. Deprotection followed with the use of TBAF in THF affording triol **47** and diol **46** in excellent yields.

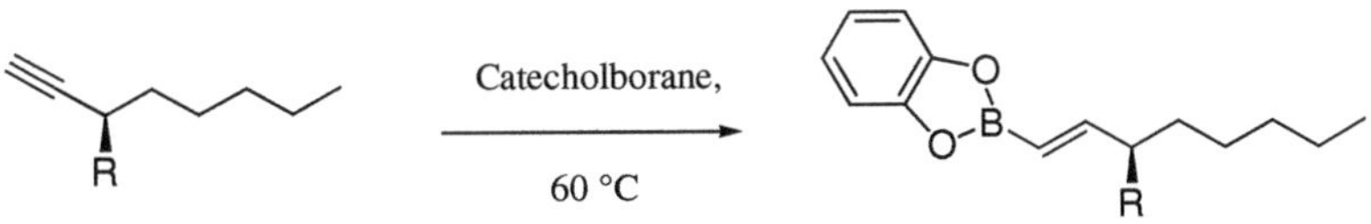

53 R = H
54 R = OTBS

55 R = H
56 R = OTBS

Scheme 2.12 Synthesis of key intermediates **55** and **56** [4]

52 + 55

52 + 56

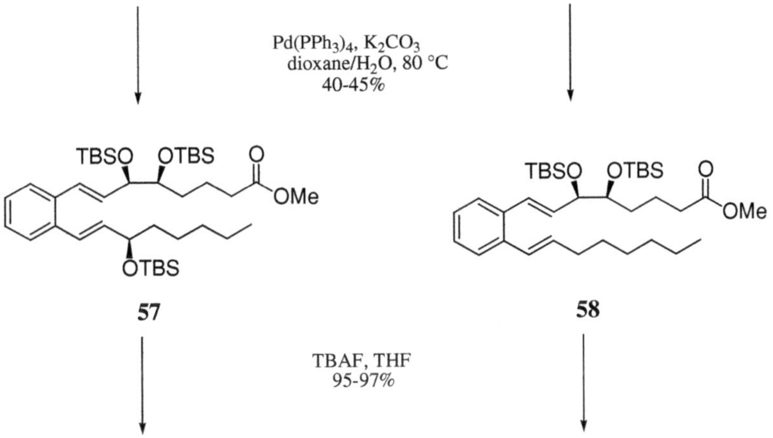

57

58

47

46

Scheme 2.13 Synthesis of key intermediates **47** and **46** [4]

The same authors also described an interesting and alternative generation of **47** involving a novel and time-conserving one pot boronic acid Heck-type coupling, Scheme 2.14. Both alkenes **35** and **59** were prepared from their corresponding aldehyde precursors, by way of an extremely useful titanium-mediated methylenation developed by Petasis and Bzowej [28]. Firstly, boronic acid **51** reacts with olefin **35** and reactivity is observed solely at the boronic acid position. In the same reaction vessel, a second Heck reaction occurs under reaction conditions reported by Jeffery [29], using $Pd(OAc)_2$, $NaHCO_3$, Bu_4NCl, PPh_3 in acetonitrile at 60°C, giving **57** in 47% yield.

The authors also described the first reported synthesis of a novel *meta*-LXA_4 analogue **48** using a related synthetic pathway starting from 3-bromophenylboronic acid **60**, Scheme 2.15. The vinyl iodide derivative **50** was coupled to **60** by way of a palladium-catalysed Suzuki–Miyaura reaction affording **61** in 70% yield. This aryl bromide **61** was further reacted in a consecutive Suzuki–Miyaura reaction with boronic ester **56**, followed by deprotection with TBAF to give the *meta*-LXA_4 analogue **48** in 42% yield over the final two steps.

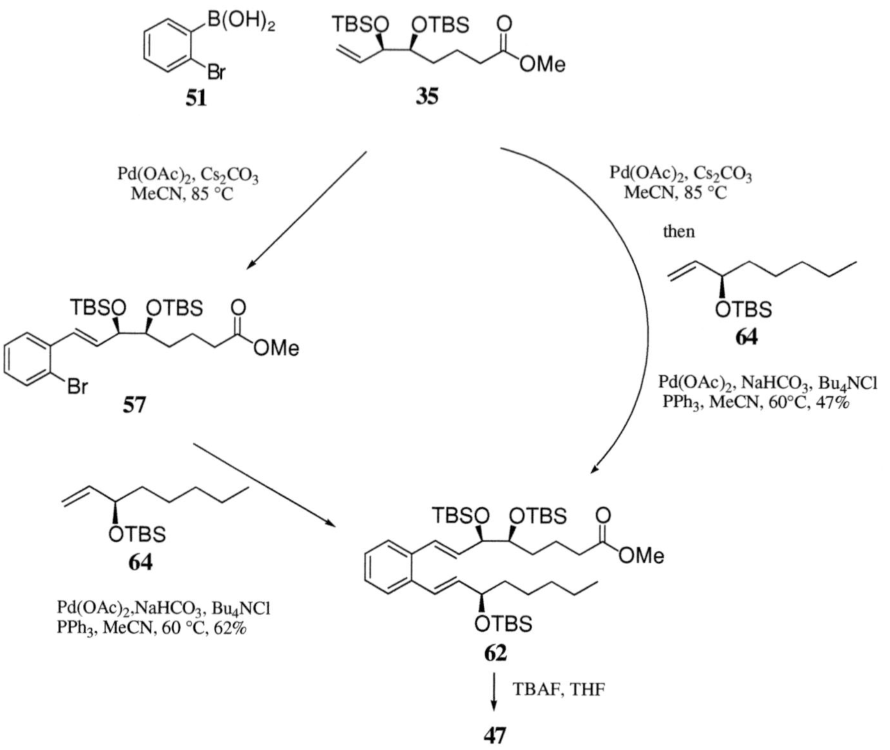

Scheme 2.14 Alternative synthesis of **47** [4]

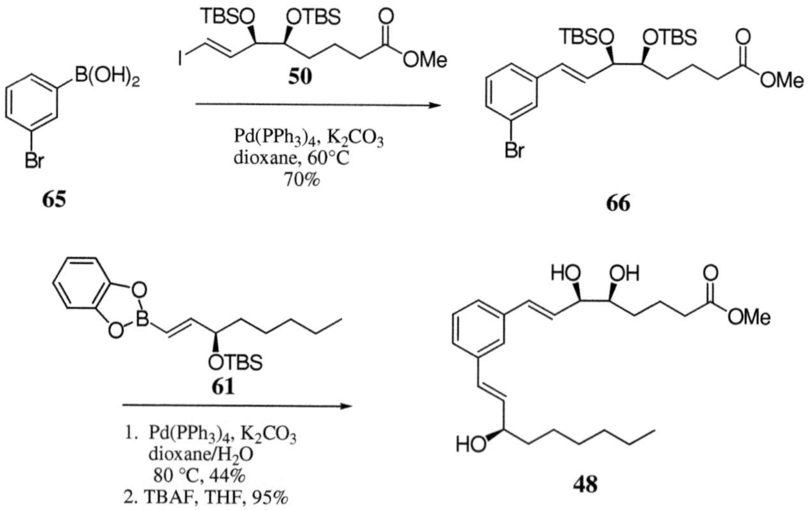

Scheme 2.15 Synthesis of a *meta*-LXA$_4$ analogue **48** [4]

The LXA$_4$ analogue **49** was prepared in order to determine the impact of increasing the chain length of the analogues on its ability to act as an agonist in the known receptor site of ALXR. The vinylboronic acid **63** was synthesised by hydroboration of the available 2-bromophenyl alkyne **62**, using the reaction conditions reported by Matteson and co-workers, Scheme 2.16 [30]. The Suzuki–Miyaura reaction of **63** with vinyl bromide **64**, prepared previously [31], gave the aryl bromide **65** in 65% yield. Conversion of this aryl bromide **65** to its pinacol boronate **66** using bis-pinacolato diboron, PdCl$_2$ (dppf) and AcOK in dimethysulfoxide at 80°C proceeded in 40% yield. Boronate **66** was coupled with vinyl iodide **50** by a Suzuki–Miyaura reaction to give the silyl-protected intermediate which was then deprotected to furnish the novel analogue **49** in 43% over the final two steps.

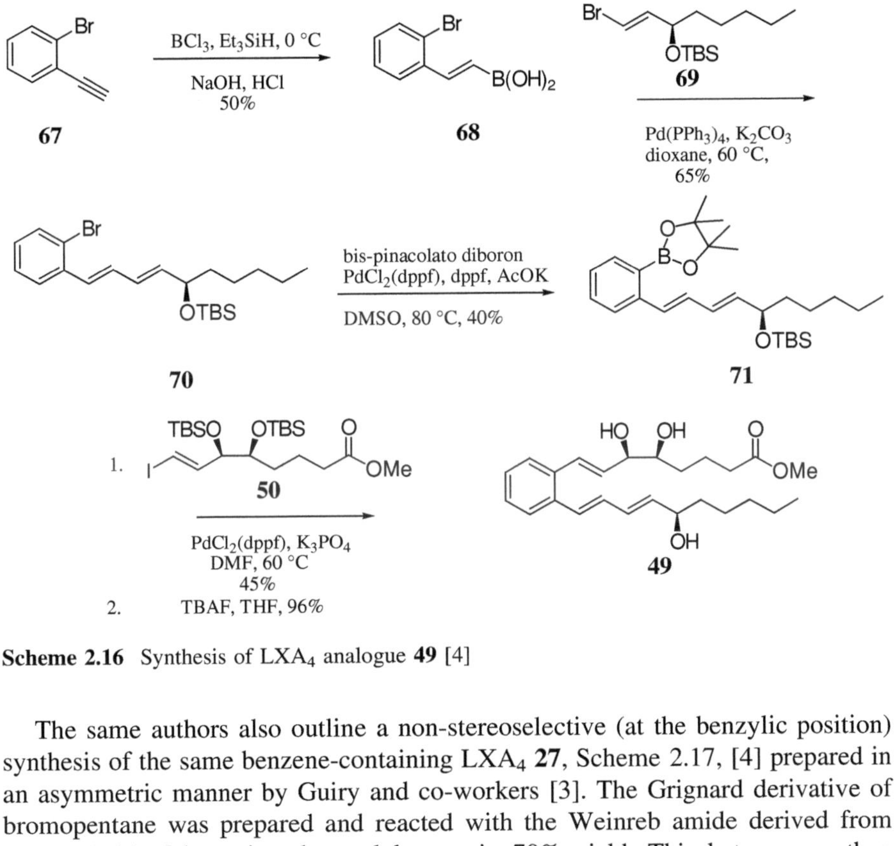

Scheme 2.16 Synthesis of LXA$_4$ analogue **49** [4]

The same authors also outline a non-stereoselective (at the benzylic position) synthesis of the same benzene-containing LXA$_4$ **27**, Scheme 2.17, [4] prepared in an asymmetric manner by Guiry and co-workers [3]. The Grignard derivative of bromopentane was prepared and reacted with the Weinreb amide derived from acid chloride **36** to give the aryl ketone in 70% yield. This ketone was then reduced using NaBH$_4$ in MeOH, followed by silyl protection to furnish **67** in high yield. Aryl bromide **67** was converted to its corresponding boronate **68** in a modest 40% yield. The *trans* olefin was constructed by the Suzuki–Miyaura coupling of boronate **68** and vinyl iodide **50** and the epimeric triol **27** was produced in 95% yield after removal of the silyl ethers.

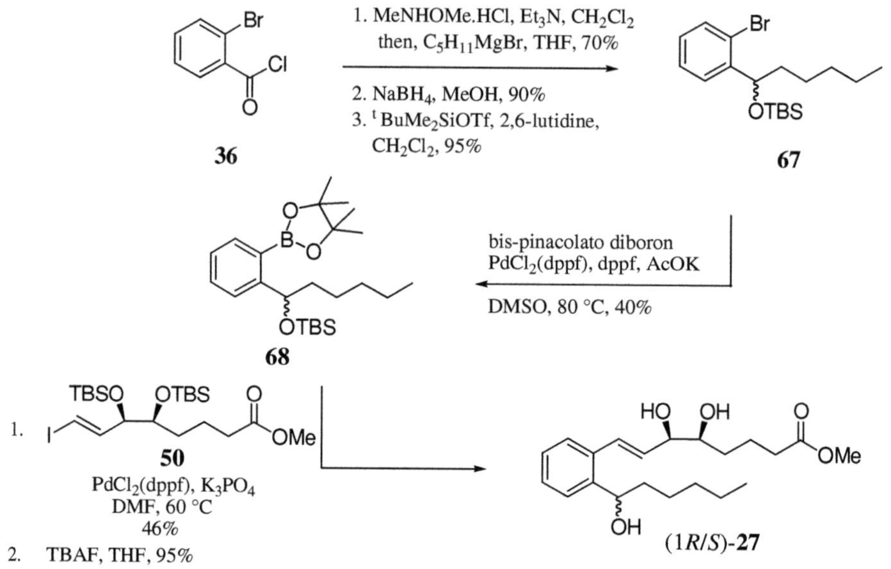

Scheme 2.17 Non-stereoselective synthesis of LXA$_4$ analogue **27** [4]

Each new stable LXA$_4$ analogue compiled by Petasis and co-workers (**27**, **46–49**) were subjected to enzymatic stability examinations in order to accurately demonstrate their resistance to rapid metabolism by recombinant eicosanoid oxido-reductase (EOR). These compounds were compared to the native LXA$_4$ **1** to determine which was metabolised the fastest, Fig. 2.12.

The deactivation was monitored by the production of the co-factor NADH. As expected, analogue **46** was the slowest to be metabolised due to the absence of a hydroxyl group on the lower chain.

These new compounds were also tested for their ability to inhibit PMN infiltration by comparison of zymosan A induced-peritonitis in mice, Fig. 2.13.

All of the above new stable analogues were found to be potentially effective anti-inflammatory agents as they increased the inhibition of PMN by up to 32% in the case of **47**. This level of activity is significant as LX and their analogues possess comparable potency to current non-steroidal anti-inflammatory drugs on the market. For example, the anti-inflammatory drug indomethacin **69**, Fig. 2.14, reduces PMN infiltration by 35–40% in the same model of peritonitis [32].

Further to this, the aromatic analogue **47** displayed therapeutic ability to reduce PMN infiltration in murine hind-limb ischemia-induced lung injury, comparable to synthetic analogues that lack the additional benzene ring moiety [24, 25]. Compound **47** was also shown to regulate the production of important cytokines and chemokines known to be fundamental in the inflammatory process [33, 34]. A decrease in MIP-2, TNF-α, and IFN-γ was observed and no effect was observed on the levels of RANTES or SDF-1.

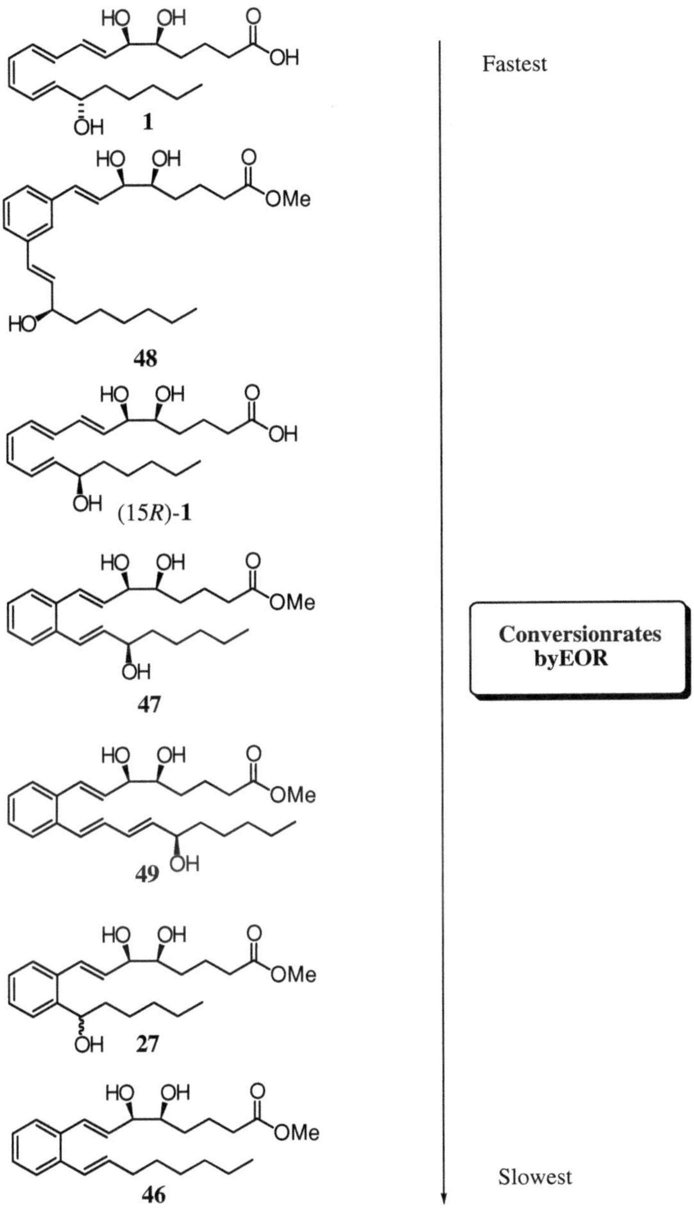

Fig. 2.12 Enzymatic metabolism by eicosanoid oxido-reductase [4]

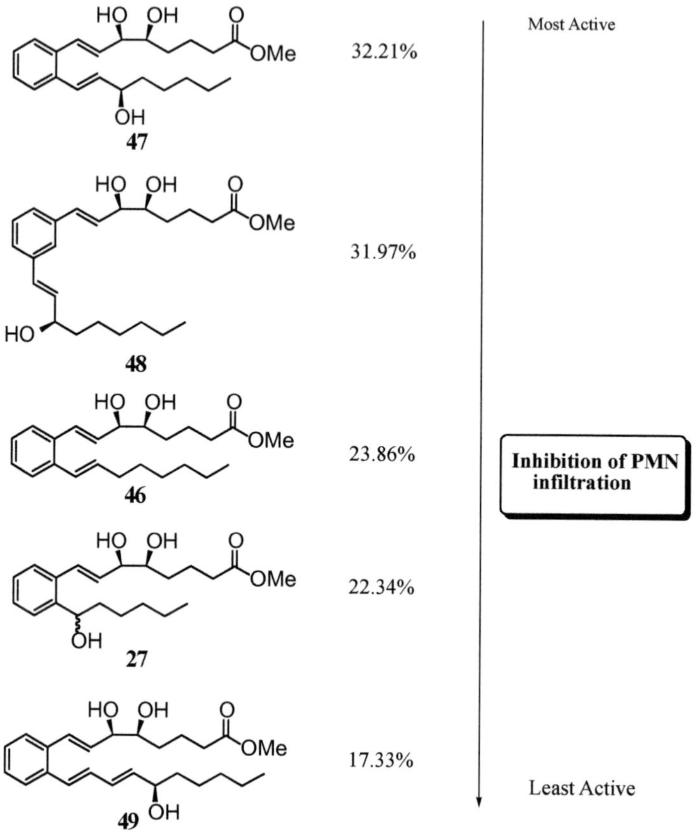

Fig. 2.13 Activity of stable analogues to inhibit PMN infiltration in vivo [4]

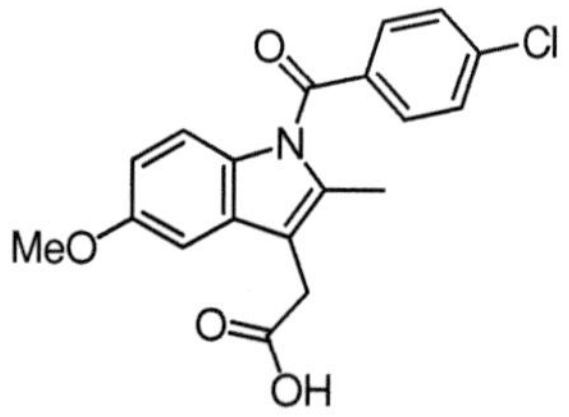

Fig. 2.14 Current non-steroidal anti-inflammatory drug indomethacin **69** [32]

2.5 (C) Structural Modifications of the Upper Chain

Although the Lipoxin receptor target has been sequenced [22], the tertiary structure has not been determined to date. Therefore, any extension and/or structural modifications of the upper chains could potentially lead to some attractive

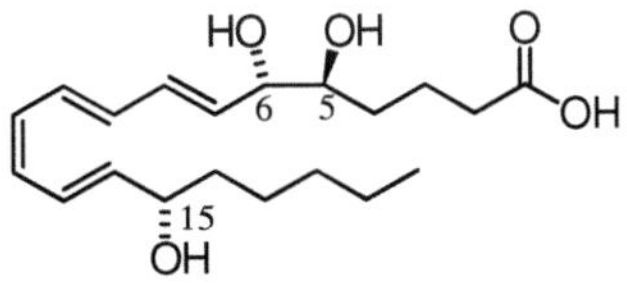

(6S)-**1**

Fig. 2.15 Inversion of stereocentre at C_6 [32]

biological findings, as chemical alterations of the lower chain have proven to be extremely advantageous in the previously reported *para*-fluorophenoxy Lipoxin analogue **6**. Structural modification of the top chain is a less researched area as the stereocentres at the hydroxyl groups are essential for bioactivity. The conversion of the stereocentre at C_6 to the corresponding (*S*) stereocentre results in a complete loss of activity, Fig. 2.15 [35].

The C_5 and C_6 hydroxyl groups have displayed resistance to enzymatic metabolism by EOR, therefore rendering this an undesirable part of the Lipoxin structure to alter. However, Guilford and co-workers discovered β-oxidation can occur at C_3 in the *para*-fluorophenoxy analogue **6**, Scheme 2.18 [5].

Stability experiments carried out on plasma samples by Guilford and co-workers revealed an unexpected result. The *para*-fluorophenoxy analogue **6** was converted

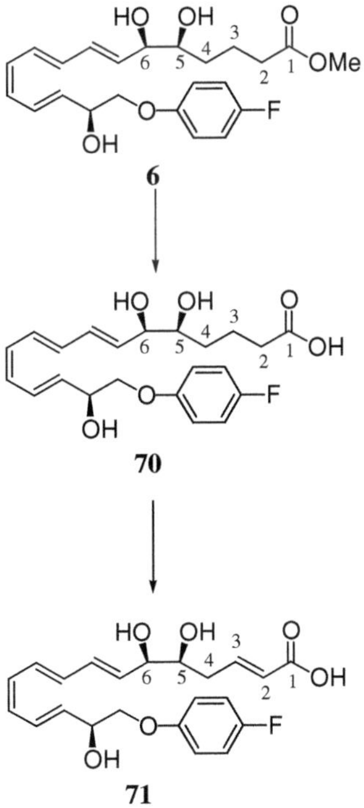

Scheme 2.18 In vivo metabolism of *para*-fluorophenoxy analogue **6** [5]

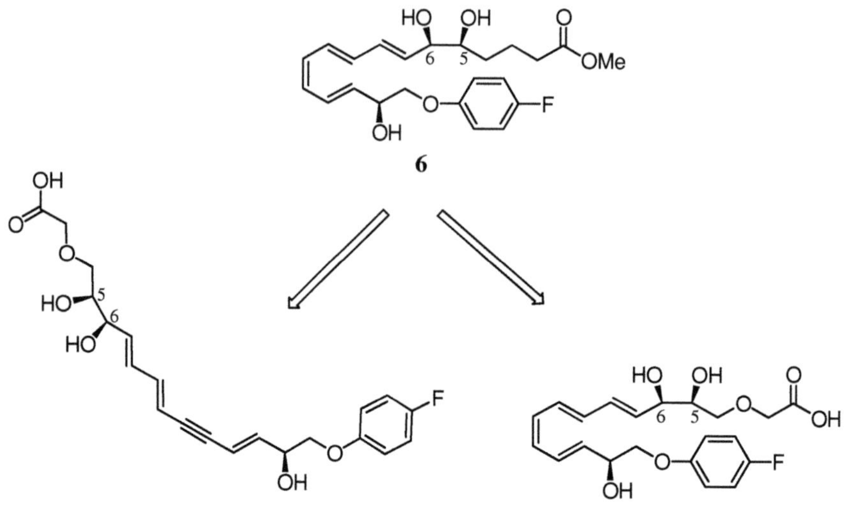

Fig. 2.16 Design of stable analogues **72** and **73** by preventing β-oxidation [5]

into the corresponding acid **70** followed immediately by β-oxidation to furnish the 2,3-dehydro analogue **71**. The assignment of this structure was aided with direct comparisons to the lipid metabolisms of the prostaglandin and the leukotriene pathways previously reported in the literature [36, 37]. With these findings in hand, Guilford designed and synthesised two new LXA$_4$ analogues (**72** and **73**) by directly replacing the CH$_2$ group at C$_3$ with an oxygen to prevent this β-oxidation and hence proposed to enhance the metabolic and chemical stability, Fig. 2.16 [5]. The design of these analogues combines the useful strategy of domain modifications (**C**) and (**A**), Fig. 2.1, modifications the upper and lower chains.

The seleoselective synthesis of **72** and **73** relies upon a Wittig reaction of a known enyne reagent [38], a palladium-catalysed Sonogashira coupling reaction and an activated zinc reduction of an alkyne. A successful chiral pool strategy was utilised in order to achieve the correct stereochemistry at C$_5$ and C$_6$ as key intermediates for the Sonogashira coupling reaction were obtained from *L*- Rhamnose **74**, Scheme 2.19.

L-Rhamnose **74** was reacted with sulfuric acid, copper sulfate and cyclohexa-none at room temperature for 16 h to afford the corresponding protected cyclohexylidene ketal **75** in 57% yield. This was reduced using NaBH$_4$ in methanol to give the triol **76** in 88% yield. Phase transfer conditions were employed to prepare the required ester which was converted into the corresponding aldehyde **77** in 92% yield using sodium metaperiodate in a mixture of water and acetone. A Wittig coupling of aldehyde **77** and the protected alkyne **78** yielded a 2:1 of mixture of *E,E* and *E,Z* isomers as determined by ^{1}H NMR spectroscopic analysis. This mixture was dissolved in dichloromethane and treated with iodine to give the required

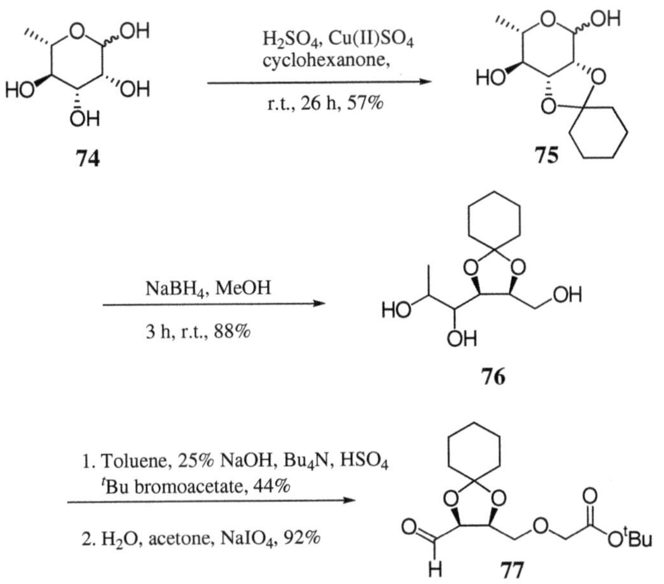

Scheme 2.19 Synthesis of key intermediate **77** [5]

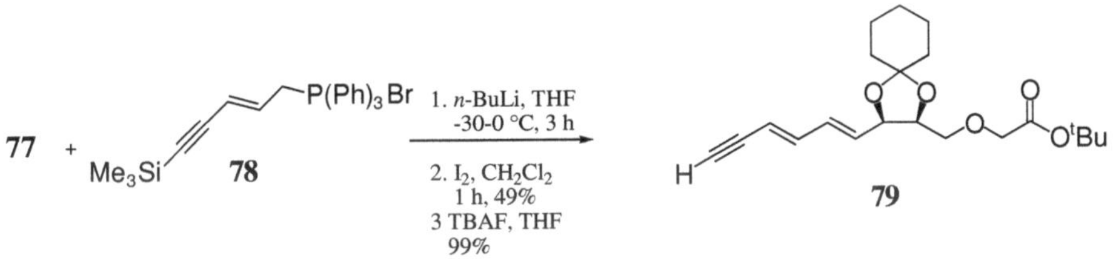

Scheme 2.20 Synthesis of key intermediate **79** [5]

protected *E,E*-dienyne in 49% yield, Scheme 2.20. This was further deprotected
using TBAF in THF giving the required terminal alkyne **79** in 99% yield.

The synthesis of the Sonogashira coupling partner **83**, Scheme 2.21, proceeded
with the conversion of carboxylic acid **80** into its acid chloride by treatment with
oxalyl chloride and a catalytic amount of DMF, followed by direct preparation of
the Weinreb amide.

This amide was treated with a solution of ethynylmagnesium bromide to furnish
the target ketone **81** in 59% yield over three steps. Ketone **81** was reduced using
R-Alpine-Borane although with a modest *ee* value of between 60 and 70%. This
problem was overcome by the conversion of the alcohol to its dinitrobenzoyl
derivative followed by a recrystallization to give *ee* values greater than 98%. This
ester was deprotected using K_2CO_3 in MeOH, followed by bromination using NBS

and silver nitrate to form the chiral alcohol **82** in 79% yield over the final two steps. Reduction of the **82** using lithium aluminium hydride and aluminium chloride gave the vinyl bromide **83**, the substrate for a subsequent Sonogashira coupling reaction, Scheme 2.21.

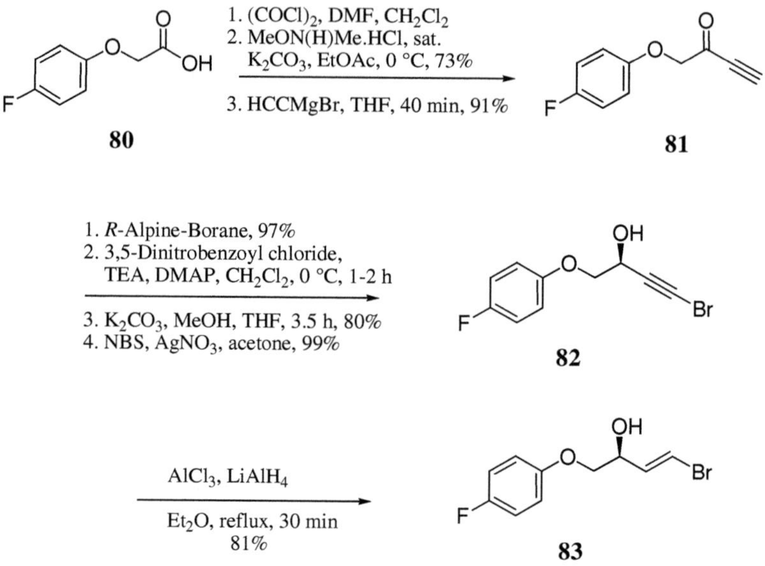

Scheme 2.21 Synthesis of vinyl bromide **83** for Sonogashira coupling

The Sonogashira coupling of **83** and **79** gave the required alkyne in 50% yield, Scheme 2.22. Cleavage of the acetal protecting group with AcOH gave diol **84** in 58% yield. Diol **84** was hydrolysed under basic conditions affording **72** in 58% yield. Reduction using activated zinc, followed by hydrolysis furnished **73** in a low 30% yield.

The natural LX along with stable analogues provide anti-inflammatory benefits in several models of induced skin inflammation [39]. With this information in hand, β-oxidation resistant analogues **72** and **73** were analysed in a calcium ionophore model topically applied to the mouse ear skin. This study revealed comparable potency to the native analogues, by inhibiting edema formation along with a decrease in neutrophil and granulocyte infiltration. Moreover, compounds **72** and **73** have demonstrated the ability to promote the resolution of colitis induced by the hapten trinitrobenzene sulfonic acid which is a model of Crohn's disease [40, 41].

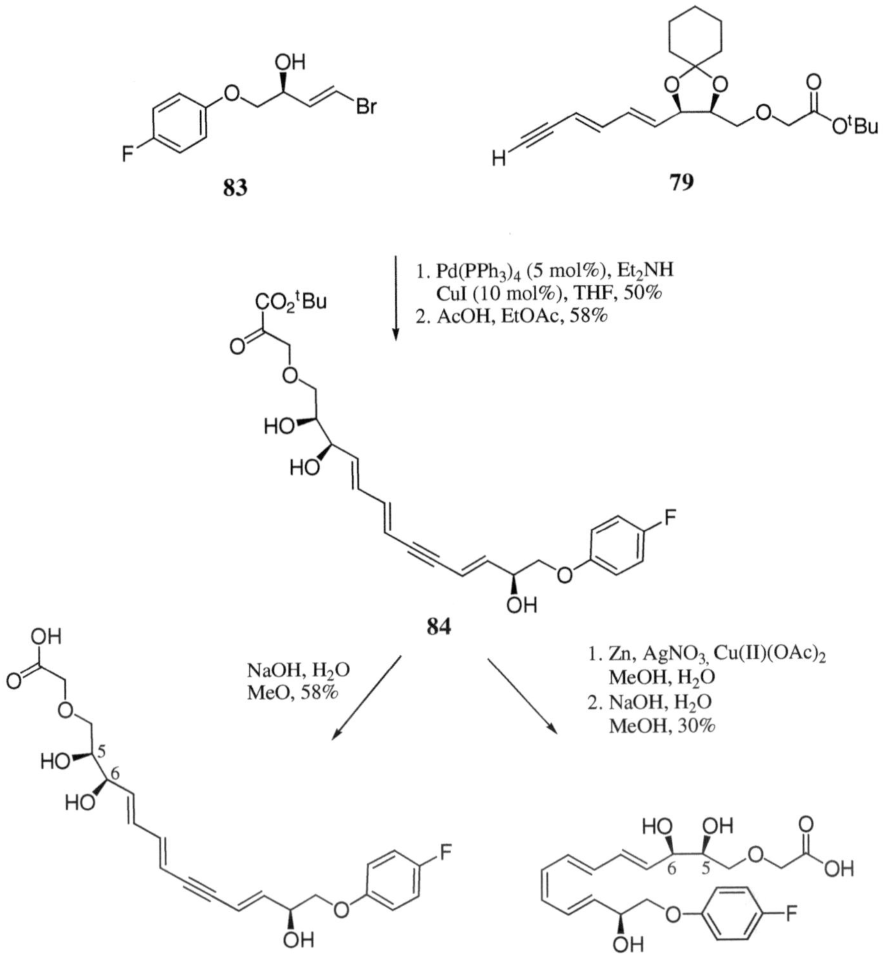

83

79

1. Pd(PPh$_3$)$_4$ (5 mol%), Et$_2$NH
   CuI (10 mol%), THF, 50%
2. AcOH, EtOAc, 58%

84

NaOH, H$_2$O
MeO, 58%

1. Zn, AgNO$_3$, Cu(II)(OAc)$_2$
   MeOH, H$_2$O
2. NaOH, H$_2$O
   MeOH, 30%

72

73

Scheme 2.22 Synthesis of stable analogues **72** and **73** [5]

2.6 Conclusion

Modifications of three key target areas on the LX structure have resulted in the development of Lipoxin analogues displaying increased bioactivity and bioavailability compared to the native LX. The potential biological applications of these stable LX analogues have resulted in a number of efficient synthetic routes being developed for their preparation. Replacement of the C$_{15-20}$ chain by cyclohexyl- and phenoxy-groups and later the further derivatisation of these analogues with fluoro-groups, gave rise to compounds which showed increased biostability and displayed potential anti-cancer properties. Phillips and Petasis pioneered the research involving stabilisation of this key C_{15-20} chain. Modification of the triene structure which is present in the native LX has been an active area of research. Incorporation of benzene or a heteroaromatic ring in place of this triene structure has had a number of enhanced properties, including stability towards enzymatic decomposition. Guiry and co-workers reported the first stereocontrolled synthesis of a benzene-containing analogue and found that it enhanced the phagocytosis of PMN by macrophages. Guiry and co-workers later published the synthesis of a novel analogue, where the triene had been replaced by a pyridine ring. They found that both epimers displayed potent anti-inflammatory characteristics. There have been fewer reports of structural modifications of the upper chain of the LX, mainly due to the importance of retaining the hydroxyl groups in order to maintain bio-activity. Guilford incorporated oxygen into the upper chain, replacing the β-CH_2 group. This resulted in an analogue that displayed resistance to β-oxidation, leading to heightened metabolic and chemical stability. This derivative also showed potential in the treatment of Crohn's disease.

This chapter reports a concise review of the synthetic and biological developments of novel stable Lipoxin analogues. The major and noteworthy synthetic obstacles and achievements were outlined and discussed. There is an on-going effort to provide novel therapeutic agents to combat an array of inflammatory diseases and it is hoped that this timely review will help to stimulate the design and biological evaluation of novel Lipoxin analogues.

References

1. Collins PW, Djuric SW (1993) Chem Rev 93:1533
2. Phillips ED, Chang H-F, Holmquist CR, McCauley JP (2003) Bioorg Med Chem Lett 13:3223
3. O'Sullivan TP, Vallin KSA, Shah STA, Fakhry J, Maderna P, Scannell M, Sampaio ALF, Perretti M, Godson C, Guiry PJ (2007) J Med Chem 50:5894
4. Petasis NA, Keledjian R, Sun Y-P, Nagulapalli KC, Tjonahen E, Yang R, Serhan CN (2008) Bioorg Med Chem Lett 18:1382
5. Guilford WJ, Bauman JG, Skuballa W, Bauer S, Wei GP, Davey D, Schaefer C, Mallari C, Terkelsen J, Tseng J-L, Shen J, Subramanyam B, Schottelius AJ, Parkinson JF (2004) J Med Chem 47:2157
6. Nicolaou KC, Ramphal JY, Petasis NA, Serhan CN (1991) Angew Chem Int Ed Engl 30:1100
7. Petasis NA, Akritopoulou-Zanze I, Fokin VV, Bernasconi G, Keledjian R, Yang R, Uddin J, Nagulapalli KC, Serhan CN (2005) Prostaglandins Leukot Essent Fat Acids 73:301
8. Serhan CN, Maddox JF, Petasis NA, Akritopoulou-Zanze I, Papayianni A, Brady HR, Colgan SP, Madara JL (1995) Biochemistry 34:14609–14615
9. Webber SE, Veale CA, Nicolaou KC (1988) Adv Exp Med Biol 229:61
10. Clish CB, O'Brien JA, Gronert K, Stahl GL, Petasis NA, Serhan CN (1999) Proc Natl Acad Sci U S A 96:8247
11. Fierro IM, Kutok JL, Serhan CN (2002) J Pharmacol Exp Ther 300:385
12. Rodríguez AR, Spur BW (2001) Tetrahedron Lett 42:6057

13. Nicolaou KC, Veale CA, Webber SE, Katerinopoulos H (1985) J Am Chem Soc 107:7515
14. Boland W, Schroer N, Sieler C, Feigel M (1987) Helv Chim Acta 70:1025
15. Takano T, Fiore S, Maddox JF, Brady HR, Petasis NA, Serhan CN (1997) J Exp Med 185:1693
16. Duffy CD, Maderna P, McCarthy C, Loscher CE, Godson C, Guiry PJ (2010) Chem Med Chem 5:517
17. Singh S, Duffy CD, Shah STA, Guiry PJ (2008) J Org Chem 7:6429
18. Singh S, Guiry PJ (2009) Eur J Org Chem
19. Singh S, Guiry PJ (2009) J Org Chem 74:5758
20. Maderna P, Cottell DC, Berlasconi G, Petasis NA, Brady HR, Godson C (2002) Am J Pathol 160:2275
21. Maddox JF, Serhan CN (1996) J Exp Med 183:137
22. Maddox JF, Hachicha M, Takano T, Petasis NA, Fokin VV, Serhan CN (1997) J Biol Chem 272:6972
23. Chiang N, Serhan C, Dahlén S-E, Drazen JM, Hay DWP, Rovati GN, Shimizu T, Yokomiao T, Brink C (2006) Pharmacol Rev 58:463
24. Bannenberg G, Moussignac R-L, Gronert K, Devchand PR, Schmidt BA, Guilford WJ, Bauman JG, Subramanyam B, Perez HD, Parkinson JF, Serhan CN (2004) Br J Pharmacol 143:43
25. Sun Y-P, Tjonahen E, Keledjian R, Zhu M, Yang R, Recchiuti A, Pillai PS, Petasis NA, Serhan CN (2009) Prostaglandins Leukot Essent Fat Acids 81:357
26. Nicolaou KC, Webber SE (1984) J Am Chem Soc 106:5734
27. Midland MM, McDowell DC, Hatch RL, Tramontano A (1980) J Am Chem Soc 102:867
28. Petasis NA, Bzowej EI (1990) J Am Chem Soc 112:6392
29. Jeffery T (1996) Tetrahedron 52:10113
30. Soundararajan R, Matteson DS (1990) J Org Chem 55:2274
31. Petasis NA, Zavialov IA (1996) Tetrahedron Lett 37:567
32. Hong S, Gronert K, Devchand PR, Moussignac R-L, Serhan CN (2003) J Biol Chem 278:14677
33. Serhan CN, Hong S, Gronert K, Colgan SP, Devchand PR, Mirick G, Moussignac R-L (2002) J Exp Med 196:1025
34. Sodin-Semrl S, Taddeo B, Tseng D, Varga J, Fiore S (2000) J Immunol 164:2660
35. Samuelsson B, Dahlen SE, Lindgren JA, Rouzer CA, Serhan CN (1987) Science 237:1171
36. Ellis CK, Smigel MD, Oates JA, Oelz O, Sweetman BJ (1979) J Biol Chem 254:4152
37. Guindon Y, Delorme D, Lau CK, Zamboni R (1988) J Org Chem 53:267
38. Robinson RA, Clark JS, Holmes AB (1993) J Am Chem Soc 115:10400
39. Schottelius AJ, Giesen C, Asadullah K, Fierro IM, Colgan SP, Bauman J, Guilford W, Perez HD, Parkinson JF (2002) J Immunol 169:7063
40. Fiorucci S, Wallace JL, Mencarelli A, Distrutti E, Rizzo G, Farneti S, Morelli A, Tseng J-L, Suramanyam B, Guilford WJ, Parkinson JF (2004) Proc Natl Acad Sci U S A 101:15736
41. Guilford WJ, Parkinson JF (2005) Prostaglandins Leukot Essent Fat Acids 73:245

Chapter 3
Synthesis of Heck Coupling Partner for the Preparation of Heteroaromatic Lipoxin A$_4$ Analogues

3.1 Introduction

Our research group recently developed a short and efficient synthetic route for the preparation of novel stable benzene-containing Lipoxin A$_4$ and Lipoxin B$_4$ analogues, reviewed in Chap. 2 [1]. One of the key synthetic steps relies on the construction of a *trans* double bond via a palladium-catalysed Heck reaction, a reliable and powerful method for the assembly of this class of alkene, Scheme 3.1 [2, 3].

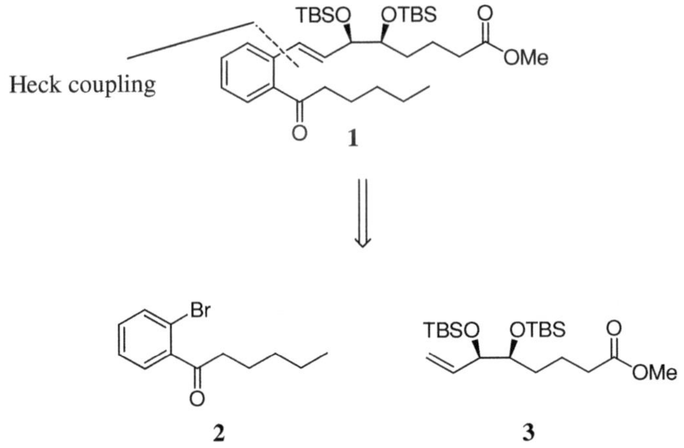

Scheme 3.1 Reterosynthesis of benzene-containing LXA$_4$

Compound **1** is a key intermediate in the synthesis of stable benzene-containing LXA$_4$ analogues. The preparation of this key intermediate involves the coupling of aryl bromide **2** and terminal olefin **3**. Olefin **3** can be considered an important Heck coupling partner as this intermediate can potentially be reacted with a variety of aryl or heteroaryl halides. Our continued interest in the preparation of stable

C. Duffy, *Heteroaromatic Lipoxin A$_4$ Analogues*, Springer Theses,
DOI: 10.1007/978-3-642-24632-6_3, © Springer-Verlag Berlin Heidelberg 2012

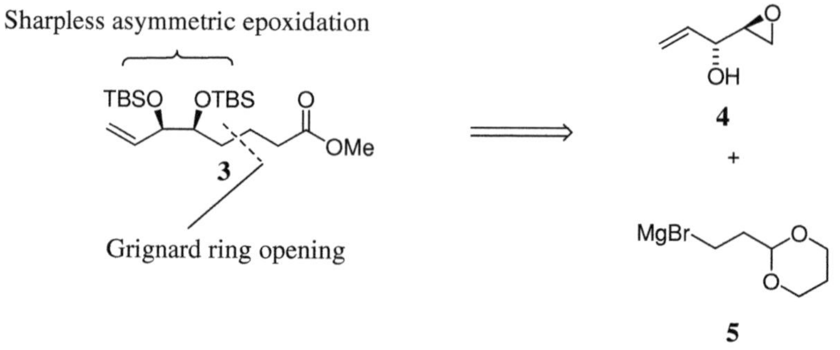

Sharpless asymmetric epoxidation

Fig. 3.1 Reterosynthetic analysis of key intermediate **3**

bioactive compounds has led us to design a series of heteroaryl Lipoxin analogues, which will be discussed in Chaps. 4, 5 and 6. This chapter will describe the preparation of key intermediate **3**, whose synthesis relies on a Sharpless asymmetric epoxidation, a Grignard reaction and a novel one-pot zirconium tetrachloride-catalysed deprotection/transesterification protocol (Fig. 3.1).

3.2 Synthesis of Key Intermediate for Heck Coupling Reaction

The Sharpless asymmetric epoxidation of allylic alcohols is one of the most widely used reactions in natural product synthesis owing to its high enantioselectivity and excellent yields [4, 5]. The epoxidation of divinylcarbinol **6** was carried out using the procedure reported by Wong and Romero which furnished epoxide **4** in 80% yield and >99% *ee*, Scheme 3.2 [6].

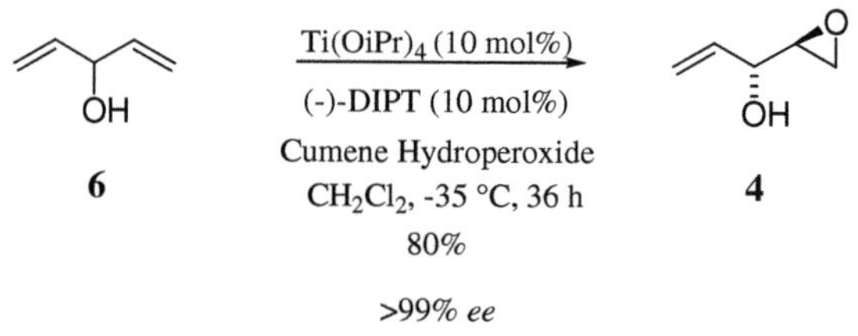

Scheme 3.2 Asymmetric synthesis of epoxide **4**

This method has the advantage of using technical grade cumene hydroperoxide rather than high purity *tert*-butyl hydroperoxide. We also noticed a dramatic increase in yield from 55 to 80% when using cumene hydroperoxide rather than *tert*-butyl hydroperoxide.

The formation of the epoxide **4** was evident from the ^{1}H NMR spectrum, Fig. 3.2. A multiplet was observed at 2.79 ppm integrating for two protons

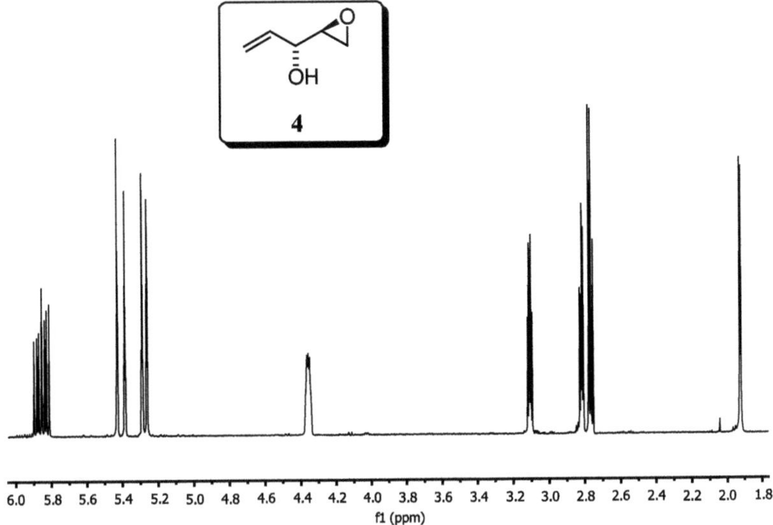

Fig. 3.2 400 MHz ^{1}H NMR spectrum of epoxide **4**

corresponding to the terminal epoxide CH_2. This was accompanied by a multiplet at 4.35 ppm integrating for the CH proton bonded to the hydroxyl group.

A characteristic hydroxyl stretch was observed at 3,400 cm^1 in the IR spectrum. The optical rotation value obtained for epoxide 4 was consistent to a previously reported literature value [5]. The enantiomeric excess was determined by chiral GC of the acetylated epoxide as attempts to separate the racemic epoxide were unsuccessful. The enantiomeric excess achieved was greater than 99%.

The next step in the synthesis required the ring opening of this chiral epoxide **4**. This was accomplished with the use of a Grignard reagent, an extremely useful method for the formation of carbon–carbon bonds [7–9]. Addition of the Grignard derivative of **5** to the epoxide **4** in the presence of a catalytic amount of copper iodide furnished *syn*-diol **7** in 75% yield, Scheme 3.3.

Scheme 3.3   Synthesis of *syn*-diol **6**

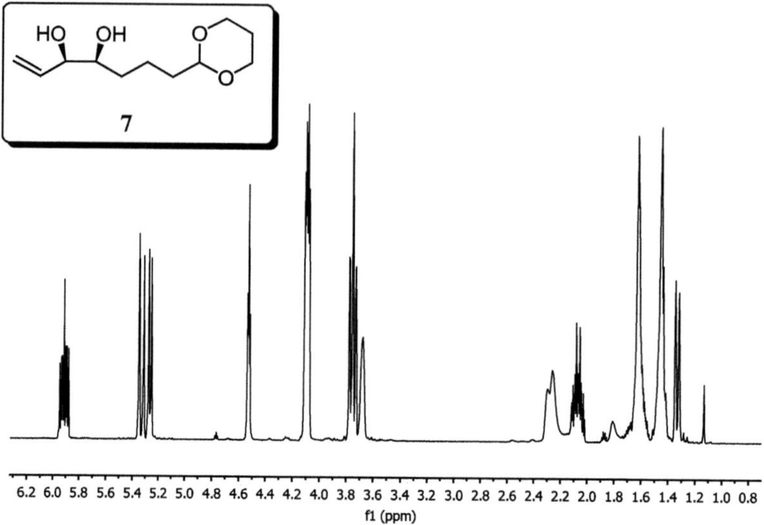

Fig. 3.3 500 MHz ^{1}H NMR spectrum of *syn*-diol **7**

Required acid-stable protecting group

Fig. 3.4 Protecting group strategy

The ^{1}H NMR spectrum of **7** showed the disappearance of the epoxide CH_2 signals at 2.79 ppm, Fig. 3.3. A triplet at 4.53 ppm, integrating for one proton, corresponding to the CH on the dioxane ring, also was sufficient evidence to suggest the product had formed. A strong broad stretch was also observed for the hydroxyl groups at 3,422 cm^{-1} in the IR spectrum.

Functional group protection and deprotection is a fundamental process in the preparation of bioactive molecules [10]. At this point in the synthesis we needed to hinder the reactivity of the hydroxyl groups by replacing them with acid-stable protection groups. This was required as the cleavage of the dioxane ring to form the carboxylic acid required acidic Jones' oxidation conditions, Fig. 3.4.

The *syn*-diol **7** was protected using acetyl chloride and pyridine in THF at room temperature affording the diacetate **8** in 90% yield, Scheme 3.4.

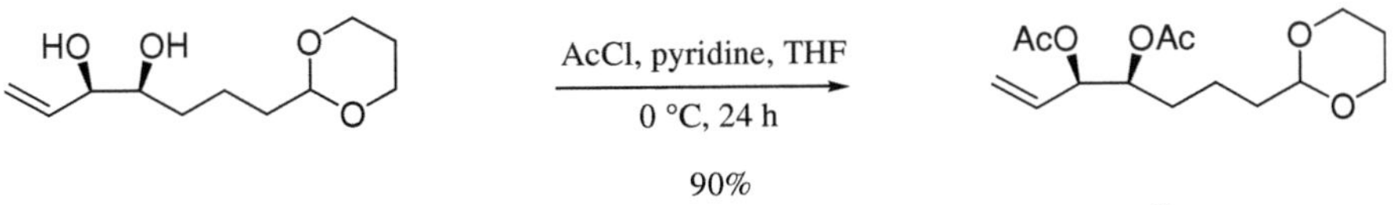

Scheme 3.4 Protection of *syn*-diol **7**

With the protection in place, we were now in a position to cleave the dioxane in order to prepare the corresponding acid **9**. Addition of excess Jones' reagent to a concentrated solution of **8** in acetone led to the cleavage of the acetal group and oxidation of the resulting aldehyde to the carboxylic acid **9** in 55% yield, Scheme 3.5.

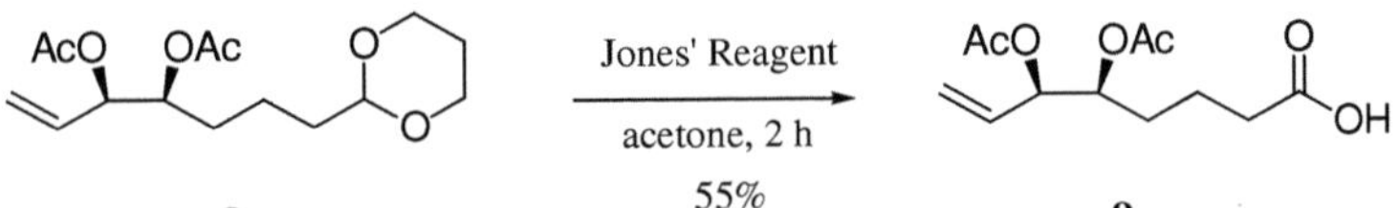

Scheme 3.5 Cleavage of the dioxane protecting group

In light of the ability of $ZrCl_4$ to catalyse a wide range of transformations [11–14] including the esterification of carboxylic acids and its potential to promote acetate deprotection, we investigated its use in a one-pot deprotection/transesterification transformation of acid **9**. We found a catalytic amount of $ZrCl_4$ (20 mol%) to be an efficient catalyst for the one-pot esterification and deprotection of acid **9** with lactone **11** formed as a minor byproduct, Scheme 3.6 [15].

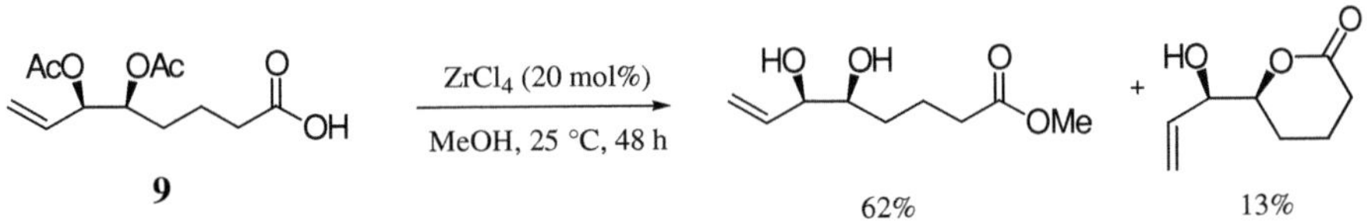

Scheme 3.6 One-pot esterification and deprotection using $ZrCl_4$ [15]

The use of $ZrCl_4$ (10–20 mol%) was also found to be sufficient to deprotect 1,3-dioxolane, bis-TBDMS and diacetate functional groups. It also promoted diol protection as the acetonide in 90% yield and acted as a transesterification catalyst for a range of esters. This methodology was also recently employed to prepare biologically important natural products such as mosquito oviposition attractant pheromones and *exo-* and *endo*-brevicomin [16, 17].

This novel one-pot zirconium tetrachloride deprotection/transesterification reaction is especially advantageous, not only as it combines two synthetic transformations in one but also eliminates the use of toxic and explosive diazomethane. In an alternative strategy carried out by previous group members, acid **9** was converted to the ester **12** in 93% yield, Scheme 3.7. Hydrolysis of the acetate groups using NaOMe in MeOH gave the required diol 10 in 70% yield.

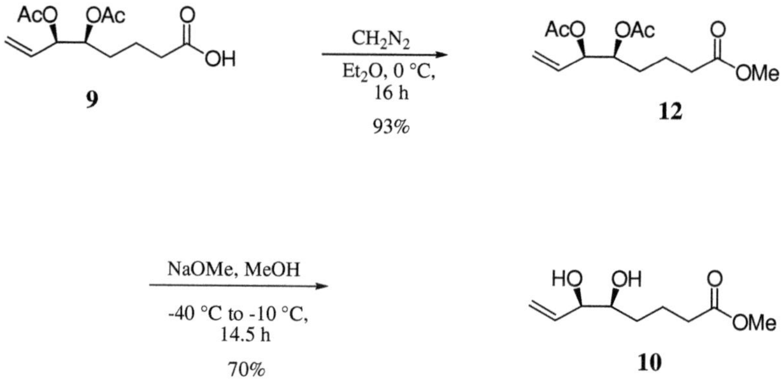

Scheme 3.7 Alternative synthesis of **10**

This diol **10** was further protected using standard conditions of *tert*-butyldimethylsilyl chloride and imidazole in DMF affording the advanced key intermediate **3** in 83% yield, Scheme 3.8 [18]. This change in protecting group strategy was necessary as the diacetate **12** was found to be an extremely poor candidate for the Heck coupling reaction [1].

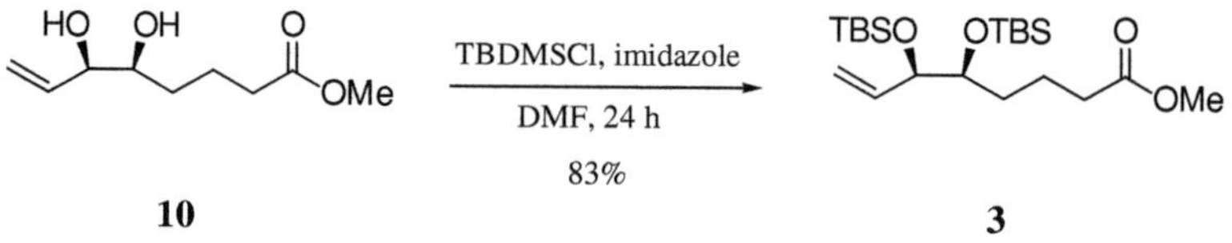

Scheme 3.8 Synthesis of advanced key intermediate **3**

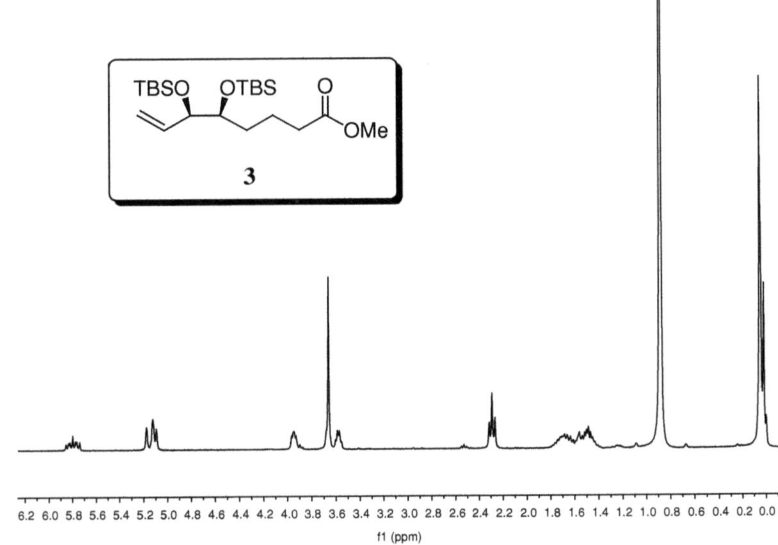

Fig. 3.5 300 MHz ^{1}H NMR spectrum of bis-silyl ether **3**

Evidence of the formation of **3** was observed in its ^{1}H NMR spectrum as the silyl ether proton signals appeared as a multiplet at 0.05 ppm and singlet at 0.88 ppm, Fig. 3.5.

3.3 Conclusion

In summary, we have developed a short and efficient synthesis for the preparation of key intermediate **3** which will be used as a Heck coupling partner in the preparation of various Lipoxin analogues, Chaps. 4, 5 and 6. This synthesis installs the required stereochemistry in the upper chain of the Lipoxin framework by way of a Sharpless asymmetric epoxidation. A Grignard ring opening reaction followed to furnish the required diol which, after a series of protection/deprotection steps, ultimately furnished the bis-silyl ether **3**. We also developed an efficient one-pot esterification and deprotection with the use of a catalytic amount of $ZrCl_4$ (20 mol%). This method was employed for the synthesis of 10, and represents an important and convenient alternative to the use of diazomethane. Our research group has also used this methodology to prepare known bioactive natural products, an emerging goal in our research laboratory [16, 17].

3.4 Experimental

3.4.1 (R)-1(S)-Oxiran-2-yl)prop-2-en-1-ol (4)

A mixture of crushed 4 Å molecular sieves (2 g) and CH_2Cl_2 (60 mL) was cooled to -35 °C and $Ti(OiPr)_4$ (1.6 g, 5.59 mmol) and (R,R)-$(-)$-DIPT (1.8 g, 7.7 mmol) were added. The mixture was stirred at -35 °C for 30 min, divinyl-carbinol 6 (5.0 g, 59.5 mmol) was added followed by cumene hydroperoxide (18.0 g, 119 mmol) over 30 min. The reaction mixture was stirred at -35 °C for 36 h. Aqueous saturated Na_2SO_4 (5 mL) was added and the mixture was diluted with Et_2O (50 mL). After the mixture was stirred at room temperature for 3 h it was filtered through a pad of Celite. The resulting yellow solution was concentrated. Excess cumene hydroperoxide was removed by silica gel chromatography (pentane/ ethyl acetate, 4:1 then neat Et_2O). The epoxide 4 was distilled (28 mmHg, 120 °C) as a colourless oil (4.7 g, 80% yield) TLC: $R_f = 0.51$ (pentane/ethyl acetate, 1:1); $[\alpha]_D^{20}$ -55.4 ($c = 1.0$, $CHCl_3$) Lit.[6] $[\alpha]_D^{25}$ -53.0 ($c = 0.73$, $CHCl_3$); ^1H NMR (400 MHz, $CDCl_3$) δ 5.86 (ddd, $J = 17.2, 10.9, 6.6$ Hz, 1H), 5.43-5.25 (m, 2H), 4.38$-$4.34 (m, 1H), 3.1 (dd, $J = 2.9, J = 3.1$ Hz, 1H), 2.83-2.75 (m, 2H), 1.93 (d, $J = 2.8$, 1H); ^{13}C NMR (125 MHz, $CDCl_3$) δ 135.6, 117.6, 70.2, 53.9, 43.6; IR (neat) (v_{max}, cm^{-1}) 3400, 2988; HRMS (ESI) Found 101.0608 $[M + H]^+$ $C_5H_9O_2$ requires 101.0603.

3.4.2 (3R, 4S)-7-[1',3']Dioxan-2'-yl-hept-1-ene-3,4-diol (7)

The Grignard derivative of 2-(2-bromoethyl)-1,3-dioxane was prepared by addition of the bromide 5 (6.2 g, 0.32 mmol) to preactivated magnesium turnings (0.72 g, 30 mmol) in THF (50 mL) followed by heating to reflux for 45 min. The solution was transferred to a 2-necked flask containing copper(I) iodide (0.381 g,

2 mmol) at −35°C and stirred for 5 min. The epoxide **4** (0.24 g, 10 mmol) in THF (5 mL) was added dropwise over 20 min and stirring was continued for a further 3 h at −35°C. Solid ammonium chloride (0.25 g) was added and the solution was stirred at room temperature for 10 min. The solvent was removed in vacuo and saturated ammonium chloride solution (25 mL) was added. The solution was extracted with ethyl acetate (6 × 15 mL) and the combined organic layers were washed with water (25 mL), brine (25 mL) and dried over magnesium sulfate. After removal of the solvent in vacuo the residue was purified by column chromatography using silica gel (pentane/ethyl acetate, 4:1 then 1:1, then neat ethyl acetate) to afford the diol **7** (8.8 g, 75%) as a pale yellow oil.(Lit.[1]) TLC: R_f = 0.12 (pentane/ethyl acetate, 1:1); $[\alpha]_D^{20}$ +3.0 (c = 0.72, $CHCl_3$); ^1H NMR (300 MHz, $CDCl_3$) δ 5.92 (ddd, J = 17.1, 10.7, 6.7 Hz, 1H), 5.37-5.24 (m, 2H), 4.53 (t, J = 4.3 Hz, 1H), 4.09–4.12 (m, 3H), 3.79–3.68 (m, 3H), 3.46 (t, J = 6.7 Hz, 2H), 2.13–1.99 (m, 2H) 1.63–1.24 (m, 6H) ppm; ^{13}C NMR (125 MHz, $CDCl_3$) δ 136.0, 117.6, 102.2, 75.8, 73.9, 66.9, 34.8, 31.7, 25.8, 20.2 ppm; IR (neat) (v_{max}, cm^{-1}) 3422, 2856, 1642, 1430; HRMS (EMS) Found 215.1228 $[M-H]^-$ $C_{11}H_{19}O_4$ requires 215.1283.

3.4.3 1-(1-Acetoxy-4-[1',3']dioxan-2'-yl-butyl)-allyl acetate (8)

Diol **7** (1.5 g, 6.9 mmol) was dissolved in THF (160 mL) to which pyridine (1.2 mL, 15.26 mmol) was added. Acetyl chloride (0.818 mL, 14.49 mmol) was added dropwise over 1 h at 0 °C and stirring was continued for an additional 16 h at room temperature. The solution was neutralised with 5% HCl solution (130 mL) and extracted with ethyl acetate (4 × 150 mL). The combined organic layers were washed with water (130 mL), brine (130 mL) and dried over magnesium sulfate. The solvent was removed in vacuo and the residue was purified by column chromatography using silica gel (pentane/ethyl acetate, 4:1) to afford the diacetate **8** (1.88 g, 90% yield) as a pale yellow oil. (Lit.[1]) TLC: R_f = 0.58 (pentane/ethyl acetate, 1:1); $[\alpha]_D^{20}$ −26.6 (c = 0.96, $CHCl_3$); ^1H NMR (300 MHz, $CDCl_3$) δ 5.77 (ddd, J = 17.1, 10.3, 6.6 Hz, 1H), 5.36–5.23 (m, 1H), 5.06-5.01 (m, 1H), 4.49 (t, J = 4.8 Hz, 1H), 4.16–4.06 (m, 4H), 3.74 (t, J = 11.3 Hz, 3H), 2.10–2.05 (m, 8H), 1.59–1.24 (m, 6H) ppm; ^{13}C NMR (125 MHz, $CDCl_3$) δ 170.8, 70.2, 132.1, 119.7, 102.2, 75.4, 73.9, 67.1, 35.1, 29.6, 26.0, 21.3 20.2 ppm; IR (neat) (v_{max}, cm^{-1}) 2852, 1741, 1646, 1226; HRMS (EMS) Found 323.1471 $[M + Na]^+$ $C_{15}H_{24}O_6$ requires 323.1471.

3.4.4 (5S,6R)-5,6-Diacetoxy-oct-7-enoic acid (9)

Acetal **8** (1.7 g, 5.9 mmol) was dissolved in acetone (2 mL) to which Jones' reagent (8.5 mL) was added over 5 min. The solution was stirred at room temperature for 1.5 h. Isopropanol (20 mL) was added and stirring was continued for a further 15 min. The mixture was filtered through a pad of silica gel and washed with ethyl acetate (130 mL). The solvent was removed in vacuo and the residue was purified by column chromatography using silica gel (pentane/ethyl acetate, 2:1) to afford the acid **9** (852 mg, 55% yield) as a brown oil. (Lit.[1]) TLC: $R_f = 0.21$ (pentane/ethyl acetate, 1:1); $[\alpha]_D^{20}$ −19.6 ($c = 0.8$, $CHCl_3$); ^{1}H NMR (300 MHz, $CDCl_3$) δ 5.76 (ddd, $J = 17.1, 10.2, 6.3$ Hz, 1H), 5.37–5.29 (m, 3H), 5.08–5.02 (br. m, 1H), 3.38 (t, $J = 5.7$ Hz, 2H), 2.09 (s, 3H) 2.07 (s, 3H), 1.72–1.62 (m, 4H); ^{13}C NMR (125 MHz, $CDCl_3$) δ 178.7, 170.8, 170.1, 131.7, 119.3, 75.1, 73.2, 33.4, 28.7, 21.0, 20.9, 20.5 ppm; IR (neat) (ν_{max}, cm^{-1}) 3230, 2964, 1845, 1741, 1712, 1644, 1600; HRMS (EMS) Found 281.0990 $[M + Na]^+$ $C_{12}H_{18}O_6Na$ requires 281.1001.

3.4.5 (5S,6R)-5,6-Diacetoxy-oct-7-enoic acid
methyl ester (12)

Using Diazomethane distillation apparatus, acid **9** (1.5 g, 5.8 mmol) was dissolved in THF (10 mL) and added dropwise over 5 min to a cooled ethereal solution of diazomethane prepared from Diazald (5.3 g, 3 equiv.). Stirring was continued for 16 h at 0 °C. The reaction was neutralised with AcOH and extracted with ethyl acetate. The solvent was removed in vacuo and the residue was purified by column chromatography using silica gel (pentane/ethyl acetate, 4:1) to afford the ester **12** (1.46 g, 93% yield) as a colourless oil. (Lit.[1]) TLC: $R_f = 0.68$ (pentane/ethyl acetate, 7:3); $[\alpha]_D^{20}$ -28.5 ($c = 0.75$, $CHCl_3$); ^{1}H NMR (300 MHz, $CDCl_3$) δ 5.78 (ddd, $J = 17.1, 10.3, 6.7$ Hz, 1H), 5.39–5.29 (m, 3H), 5.06–5.04 (br. m, 1H), 3.67 (s, 3H), 2.34 (t, $J = 6.7$ Hz, 2H), 2.07 (s, 3H), 2.06 (s, 3H), 1.73–1.61 (m, 4H) ppm; ^{13}C NMR (125 MHz, $CDCl_3$) δ 173.5, 170.6, 169.0, 131.8, 119.6, 73.2, 60.4, 51.6, 33.5,

28.8, 21.0, 21.0, 20.8 ppm; IR (neat) (ν_{max}, cm^{-1}) 3057, 2954, 1739, 1646, 1436, 1243; HRMS (EMS) Found 295.1165 [M + Na]$^{+}$ C$_{13}$H$_{20}$O$_6$Na requires 295.1158.

3.4.6 *(5S,6R)-5,6-Dihydroxy-oct-7-enoic acid methyl ester (10)*

Diacetate **12** (1.3 g, 4.7 mmol) was dissolved in anhydrous MeOH (22 mL). NaOMe (0.18 g, 3.28 mmol) in MeOH (4 mL) was added dropwise at $-40\ ^{\circ}$C over 20 min and stirring was continued for a further 14.5 h at $-10\ ^{\circ}$C. The solution was neutralised with AcOH and silica gel (2.6 g) was added to make a slurry. The solvent was removed in vacuo and the residue was purified by column chromatography using silica gel (pentane/ethyl acetate, 2:1) to afford the title compound **10** (620 mg, 70% yield) as a colourless oil. (Lit.[1]) TLC: $R_f = 0.22$ (pentane/ethyl acetate, 1:1); $[\alpha]_D^{20}$ -28.5 ($c = 0.75$, CHCl$_3$); ^{1}H NMR (300 MHz, CDCl$_3$) δ 5.92 (ddd, $J = 16.6$, 12.0, 5.7 Hz, 1H), 5.38-5.28 (m, 3H), 4.12 (br. d, 2H), 3.67 (s, 3H), 2.37 (t, $J = 6.0$ Hz, 2H), 1.26-1.85 (m, 4H) ppm; ^{13}C NMR (400 MHz, CDCl$_3$) δ 174.4, 136.2, 118.08, 76.2, 73.8, 51.8, 34.0, 31.5, 21.3 ppm; IR (neat) (ν_{max}, cm^{-1}) 3087, 2954, 1739, 1436, 1371, 1243; HRMS (EMS) Found 187.1055 [M-H]$^{-}$ C$_9$H$_{15}$O$_4$ requires 187.0970.

3.4.7 *(5S,6R)-Methyl 5,6-bis(tert-butyldimethylsilyloxy) oct-7-enoate (3)*

Diol **10** (725 mg, 3.85 mmol) and imidazole (840 mg, 12.25 mmol) were dissolved in DMF (20 mL) to which TBDMSCl (1.81 g, 12.02 mmol) was added. The solution was stirred at room temperature for 24 h. Following removal of the solvent in vacuo, the residue was purified by column chromatography using silica gel (pentane/diethyl ether, 9.5:0.5) to afford the bis-silyl ether **3** (1.35 g, 83% yield) as a colourless oil. (Lit.[1]) TLC: $R_f = 0.76$ (pentane/diethyl ether, 9.5:0.5); $[\alpha]_D^{20}$ +1.0 ($c = 0.244$, CHCl$_3$); ^{1}H NMR (300 MHz, CDCl$_3$) δ 5.79 (ddd, $J = 16.7$, 8.1, 6.7 Hz, 1H), 5.18-5.09 (m, 2H), 3.95-3.91 (br. m, 1H), 3.66 (s, 3H), 3.58-3.53 (br. m, 1H), 2.32 (t, $J = 7.3$ Hz, 2H), 1.68-1.49 (m, 4H), 0.88 (s, 18H),

0.05 (m, 12H) ppm; ^{13}C NMR (125 MHz, CDCl$_3$) δ 178.76, 143.72, 120.77, 82.52, 80.82, 56.14, 39.06, 37.40, 30.72, 25.45, 0.75, 0.58 ppm; IR (neat) (ν_{max}, cm^{-1}) 3087, 2954, 1739, 1436,1243; HRMS (ESI) Found 439.2693 [M + Na]$^+$ C$_{21}$H$_{44}$O$_4$Si$_2$Na requires 439.2676.

3.4.8 (5S,6R)-5,6-Dihydroxy-oct-7-enoic acid methyl ester (10)

Acid **9** (1.22 g, 4.7 mmol) was dissolved in dry MeOH (4 mL) to which ZrCl$_4$ (220 mg, 0.945 mmol) was added and stirring was continued for 48 h at room temperature. The MeOH was removed under high vacuum without applying heat. The resulting residue was purified using silica gel chromatography (CH$_2$Cl$_2$/ MeOH, 96:4) to afford diol **10** (545 mg, 62% yield) as a colourless oil. (Lit.[15]) TLC: R$_f$ = 0.22 (pentane/ethyl acetate, 1:1); $[\alpha]_D^{20}$ +2.5 (c = 1.0, CHCl$_3$); ^{1}H NMR (300 MHz, CDCl$_3$) δ 5.91 (ddd, J = 17.2, 10.4, 6.4 Hz, 1H), 5.36–5.25 (m, 2H) 4.13-09 (m, 2H), 3.71-3.60 (m, 1H), 3.67 (s, 3H), 2.35 (t, J = 7.3 Hz, 2H), 1.70–1.40 (m, 4H) ppm; ^{13}C NMR (100 MHz, CDCl$_3$) δ 174.2, 132.1, 117.6, 75.94, 73.6, 52.0, 33.7, 31.2, 21.1 ppm; IR (neat) (ν_{max}, cm^{-1}) 3087, 2954, 1739, 1436, 1371, 1243; HRMS (ESI) Found 211.0956 [M + Na]$^+$ C$_9$H$_{16}$O$_4$Na requires 211.0946. Lactone **11**: $[\alpha]_D^{20}$ +10.5 (c = 1.0, CHCl$_3$); ^{1}H NMR (400 MHz, CDCl$_3$) δ 5.85 (ddd, J = 16.3, 10.6, 5.6 Hz, 1H), 5.43-5.28 (m, 2H), 4.38–4.33 (m, 2H), 2.63–2.57 (s, 1H), 2.48–2,41 (m, 2H), 2.00–1.70 (m, 4H) ppm; ^{13}C NMR (100 MHz, CDCl$_3$) δ 171.6, 134.8, 117.7, 82.9, 73.5, 29.7, 21.5, 18.3 ppm; IR (neat) (ν_{max}, cm^{-1}) 3085, 2956, 1730, 1383, 1247; HRMS (ESI) Found 179.0686 [M + Na]$^+$ C$_8$H$_{12}$O$_3$Na requires 179.0684.

References

1. O' Sullivan TP, Vallin KSA, Shah STA, Fakhry J, Maderna P, Scannell M, Sampaio ALF, Perretti M, Godson C, Guiry PJ (2007) J Med Chem 50:5894
2. Heck RF, Nolley JP (1972) J Org Chem 37:2320
3. Nicolaou KC, Bulger PG, Sarlah D (2005) Angew Chem Int Ed Engl 44:4442
4. Katsuki T, Sharpless KB (1980) J Am Chem Soc 102:5974
5. Jager V, Schroter D, Koppenhoefer B (1991) Tetrahedron 47:2195
6. Romero A, Wong C (2000) J Org Chem 65:8264
7. Grignard V (1900) Compt Rend Acad Sci Paris 130:1322
8. Wothers PGN, Warren S, Clayden J (2001) Inorganic chemistry, Oxford University Press, New York
9. Parker RE, Isaacs NS (1959) Chem Rev 59:737

10. Jarowicki K, Kocienski P (1998) J Chem Soc, Perkin Trans 1(23):4005
11. Smitha G, Chandrasekhar S, Reddy SC (2008) Synth 6:829
12. Ishihara K, Nakayama M, Ohara S, Yamamoto H (2002) Sci 290:1140
13. Ishihara K, Nakayama M, Ohara S, Yamamoto H (2002) Tetrahedron 58:8179
14. Sharma GVM, Reddy KL, Sree Lakshmi P, Radha Krishna P (2004) Tetrahedron Lett 45:9229
15. Singh S, Duffy CD, Shah STA, Guiry PJ (2008) J Org Chem 7:6429
16. Singh S, Guiry PJ (2009) Eur J Org Chem 12:1896
17. Singh S, Guiry PJ (2009) J Org Chem 74:5758
18. Corey EJ, Venkateswarlu A (1972) J Am Chem Soc 94:6190

Chapter 4
Synthesis and Biological Evaluation of Pyridine-Containing Lipoxin A$_4$ Analogues

4.1 Introduction

The development of stable LXA$_4$ and LXB$_4$ analogues which show resistance to enzymatic degradation is an on-going research goal in drug development. Efforts directed towards derivatisation of the triene system in particular have been inspired by the encouraging results obtained from the biological evaluation of our novel aromatic analogues [1], along with those of Petasis et al. [2, 3]. Rational replacement of this triene system has previously led to more stable derivatives, where enzymatic degradation is suppressed.

The target receptor for Lipoxins (LX) was first reported and sequenced by Serhan in 1996 [4]. The three dimensional structure has not been determined to date. The absence of this information means that Structure Activity Relationships are an ideal means of elucidating specific ligand-binding mechanisms and therefore the most sensible approach for the design of novel bioactive compounds. In this context, we sought to further derivatise the triene system in an effort to enhance the biological effect observed with the aromatic analogues. Substitution of benzene with heteroaromatic systems, has previously proven to be a useful strategy in medicinal chemistry often resulting in an increased pharmacological profile [5, 6]. In light of this, we sought to replace the native triene moiety with a pyridine ring and evaluate the biological effect of this substitution, Fig. 4.1.

This pyridine replacement would allow for a Structure Activity Relationship study, whereby we could determine the effect of the decreased electron density of the heteroaromatic ring and how the extra heteroatom may alter its ability to accept hydrogen-bonds from the known receptor, ALXR.

An increase in bioactivity by replacing benzene with a heteroaromatic ring has previously been observed for many drugs. Bioactive compounds benefiting from this substitution include the potent antihistamine, Mepyramine and antipsychotic, Prothipendyl, Fig. 4.2 [6, 7].

C. Duffy, *Heteroaromatic Lipoxin A$_4$ Analogues*, Springer Theses,
DOI: 10.1007/978-3-642-24632-6_4, © Springer-Verlag Berlin Heidelberg 2012

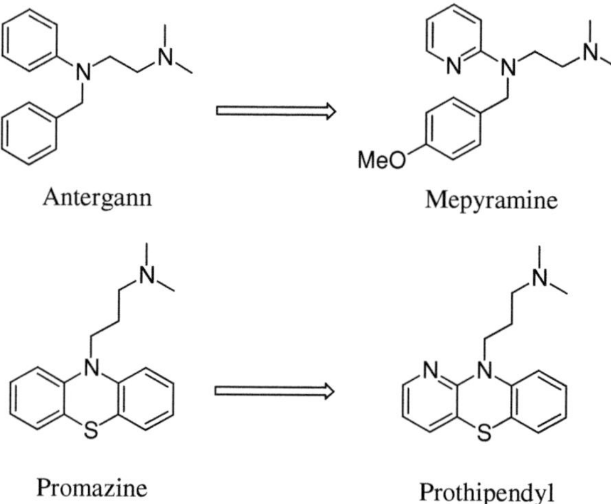

Aromatic LXA$_4$

(1*S*)-**1**
(1*R*)-**1**

Pyridine LXA$_4$

(1*S*)-**2**
(1*R*)-**2**

Fig. 4.1 Design of stable pyridine-containing LXA$_4$

Antergann

Mepyramine

Promazine

Prothipendyl

Fig. 4.2 Successful benzene/pyridine replacements in drug design

Analysis of the current best selling and most effective drugs on the market reveals that a large proportion of the compounds contain a pyridine moiety, Fig. 4.3 [8]. Pyridine-containing drugs are consequently of great interest to the pharmaceutical industry. These valuable compounds serve to treat a variety of disorders including heartburn [9], gastric reflux disease [10], diabetes [11] and duodenal ulcers [12].

The enhanced bioactivity displayed by our benzene-containing, LXA$_4$ **1**, as well as the continuous efforts to stabilise the native LX, has led us to introduce a pyridine ring into the core Lipoxin structure [13]. The aim was to further stabilise the triene against enzymatic metabolism and consequently increase the bioactivity of the heteroaromatic analogue. These heteroaromatic analogues were synthesised and evaluated for their ability to promote the clearance of apoptotic human polymorphonuclear neutrophils (PMNs). Their ability to suppress the production of key pro-inflammatory cytokines, was also examined.

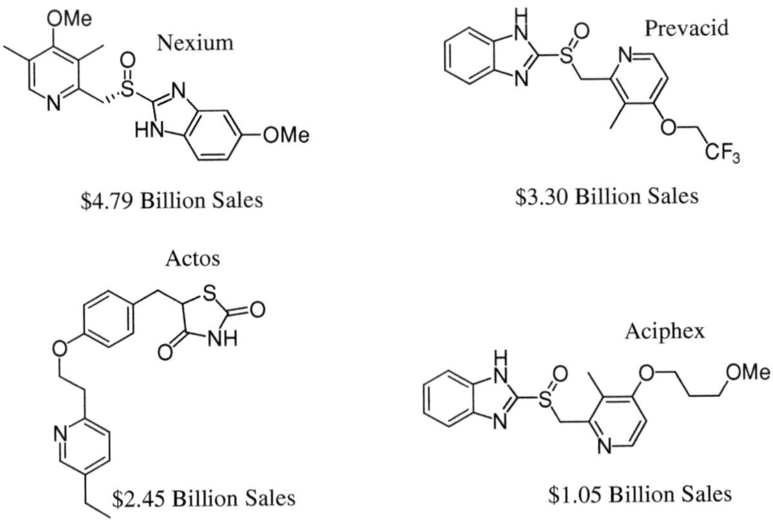

Fig. 4.3 Top selling pyridine-containing drugs in 2008 [8–12]

4.2 Retrosynthetic Analysis

The retrosynthetic analysis of the pyridine-containing LXA_4 analogue **2**, Scheme 4.1, includes an asymmetric reduction of a ketone, a palladium-catalysed Heck reaction, a Sharpless asymmetric epoxidation, discussed in detail in Chap. 2, and a regiospecific pyridine lithiation.

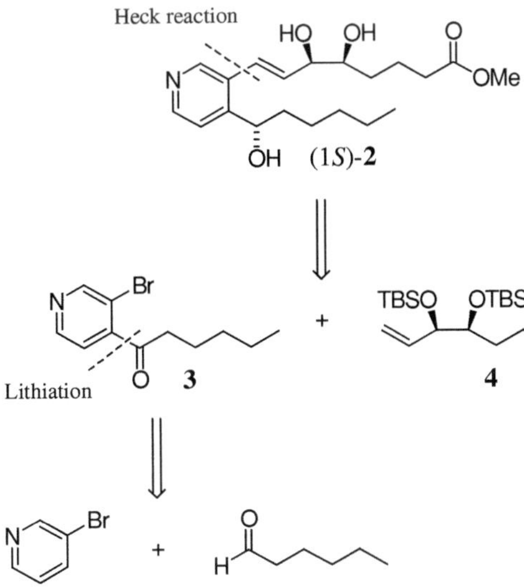

Scheme 4.1 Reterosynthetic analysis of pyridine LXA_4 (1S)-**2**

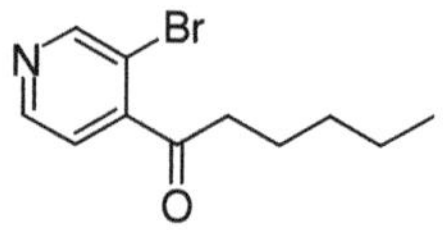

Fig. 4.4 Ketone **3** for palladium-catalysed Heck reaction [13]

4.3 Results and Discussion

The initial stage in the synthesis requires the construction of ketone **3**, Fig. 4.4, as a key intermediate for a palladium-catalysed Heck reaction.

The synthesis of this intermediate was achieved via a regiospecific pyridine lithiation of commercially available 3-bromopyridine **5**, Scheme 4.2. The procedure reported by Gribble and Saulnier [14], allows for the formation of 3,4-disubstituted pyridines in excellent yields and without the formation of unwanted side products. The key to the success of this reaction is the stability of the lithiated intermediate which is only stable for 10 min at −78 °C. The internal reaction vessel temperature must be monitored continuously as a rise in temperature above −75 °C gives rise to lithium/halogen exchange and the formation of 3,4-pyridyne. Decreasing the temperature to −100 °C prevented the formation of these unwanted by-products and gave the required intermediate **7** in 75% yield.

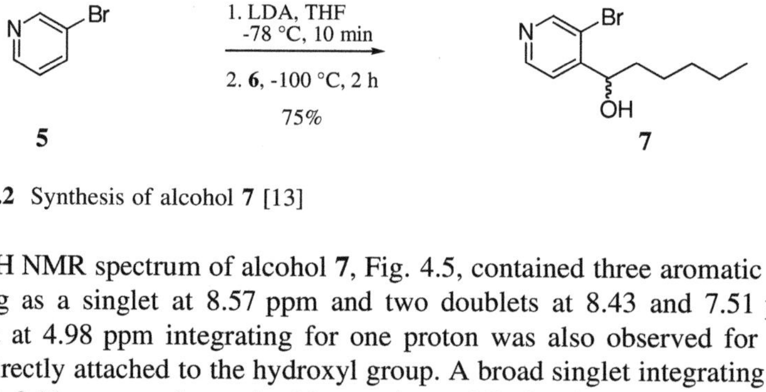

Scheme 4.2 Synthesis of alcohol **7** [13]

The ^{1}H NMR spectrum of alcohol **7**, Fig. 4.5, contained three aromatic protons appearing as a singlet at 8.57 ppm and two doublets at 8.43 and 7.51 ppm. A multiplet at 4.98 ppm integrating for one proton was also observed for the CH proton directly attached to the hydroxyl group. A broad singlet integrating for one proton at 3.11 ppm exchanged with one drop of D_2O when added to the NMR sample, identifying it as the hydroxyl proton.

We also attempted a regioselective ortho-lithiation of 3-bromopyridine **5** in order to prepare the 2,3–disubstituted pyridine LXA$_4$ analogue **8**, Fig. 4.6.

This analogue could potentially provide important information regarding the relevence of the position of the nitrogen whilst bound to the active site of the receptor target. However, all reactions carried out at −78 °C failed to produce the desired product and instead led to the formation of the 3-substituted product **9**, Scheme 4.3. This product was obtained from the quenching of 3-lithiopyridine with hexanal, due to the extremely fast lithium/halogen exchange in this system [15].

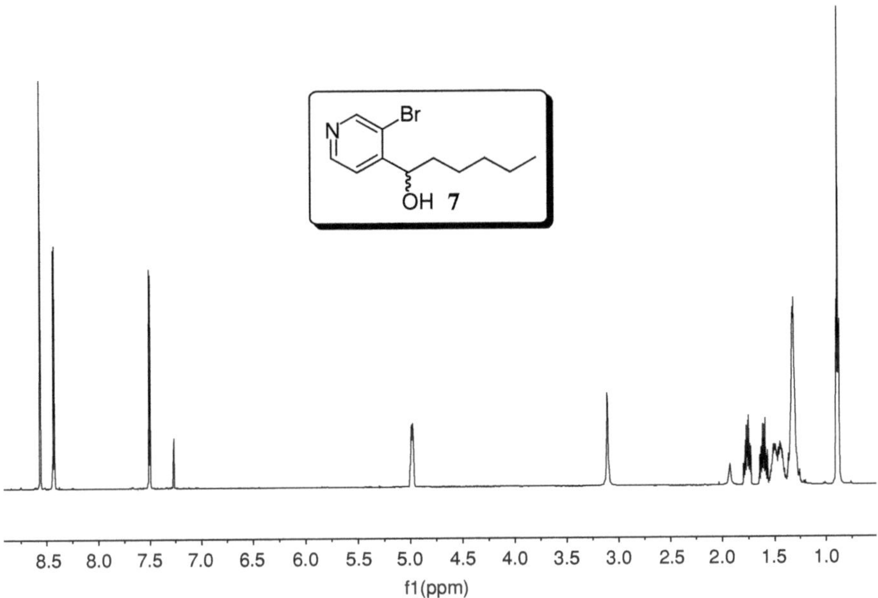

Scheme 4.3 3-Substituted product **9**

Fig. 4.5 500 MHz ^{1}H NMR spectrum of alcohol **7**

Fig. 4.6 2,3-Disubstituted pyridine LXA$_4$ analogue **8**

With the 3,4-substituted alcohol **7** in hand, we continued our synthesis in order to obtain the key intermediate **3**. The oxidation method of choice for the preparation of ketone **3** was using pyridinium chlorochromate (PCC) in the presence of glacial acetic acid, Scheme 4.4 [16, 17]. This allowed for the preparation of ketone **3** in 70% yield.

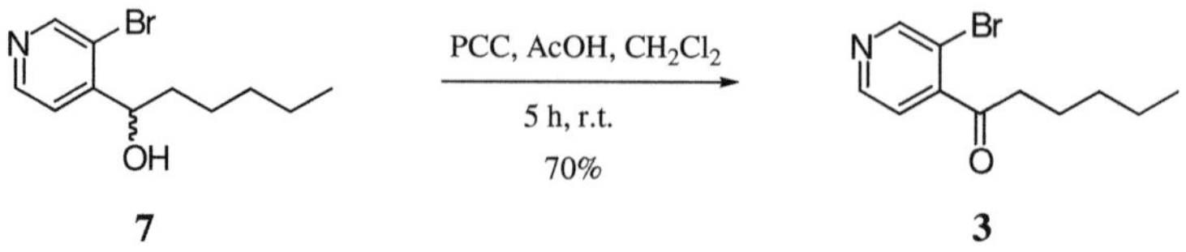

Scheme 4.4 Synthesis of ketone **3** [13]

The reaction time could be shortened by performing the reaction under microwave irradiation at 70 °C for 5 min at 150 W but gave **3** in a slightly lower 50% yield. A characteristic sharp stretch at 1710 cm^{-1} in the IR spectrum was observed for the newly formed carbonyl. This was also accompanied by a triplet in the ^{1}H NMR spectrum at 2.88 ppm integrating for the two protons of the CH_2 directly attached to the carbonyl.

An alternative and more direct preparation of ketone **3** was also attempted, Scheme 4.5. This was attempted by quenching the lithiated intermediate of 3-bromopyridine **5** with the corresponding ester or acid chloride as reported for related compounds [18]. This approach to the preparation of ketone **3** proved to be unviable as the yields obtained for this reaction ranged between 5 and 10%.

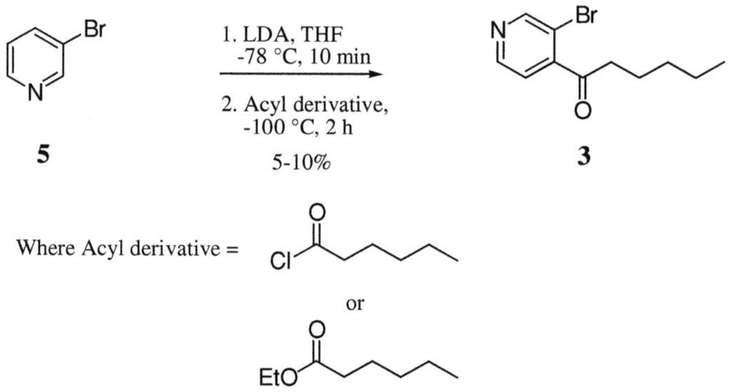

Scheme 4.5 Direct preparation of ketone **3**

With ketone **3** in hand, we attempted the preparation of the *trans* alkene using a palladium-catalysed Heck reaction. This well studied reaction is an excellent method for the formation of *trans* alkenes [19], and has been exploited in many total syntheses because of its high yield and excellent stereochemical control [20, 21]. This reaction has also been applied to many substrates on an industrial scale in the synthesis of important pharmaceutical agents [22]. For this reason, the Heck reaction was our method of choice for the coupling of ketone **3** and olefin **4**, Fig. 4.7.

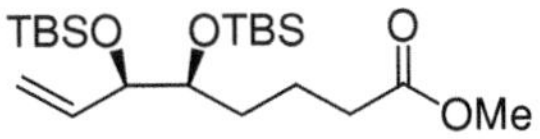

Fig. 4.7 Heck coupling partner olefin **4**

Olefin **4** was a substrate for the development of a zirconium tetrachloride-catalysed one-pot protection/deprotection synthetic methodology and its synthesis was discussed in detail in Chap. 3 [23]. Further to this, our research group employed olefin **4** as a key intermediate in the synthesis of the benzene-containing aromatic Lipoxin A$_4$ analogues, reviewed in Chap. 2 [1]. However, attempts to apply the conditions used in the latter synthesis, using palladium acetate and tri-*o*-tolylphosphine with tributylamine as the solvent and the base, only resulted in the isolation of trace amounts of the required product **10**, Scheme 4.6.

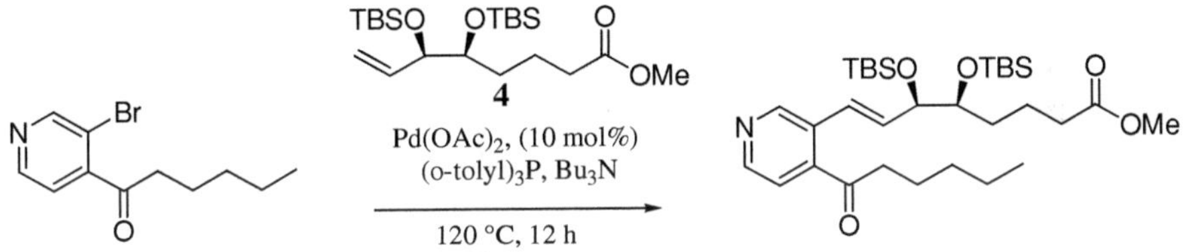

Scheme 4.6 Initial Heck coupling conditions [1]

A survey of the literature revealed alternative reaction conditions used for Heck coupling of pyridine-containing substrates [24]. These reaction conditions proved extremely successful in the synthesis of the nicotinic acetylcholine receptor intermediate **13**, Scheme 4.7.

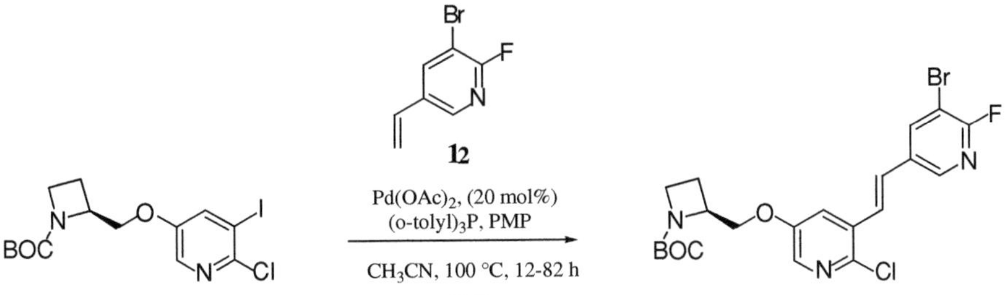

Scheme 4.7 Synthesis of nicotinic acetylcholine receptor intermediate **13** [24]

In light of this success, we employed these reaction conditions for the Heck coupling of our ketone **3** and olefin **4**. The use of palladium acetate, tri-*o*-tolyphosphine, PMP as the base in acetonitrile at 100 °C for 7 days furnished the desired product **10** in a modest 40% yield, Scheme 4.8.

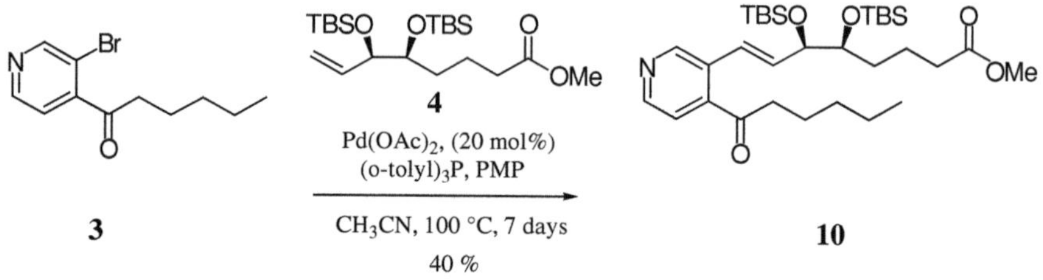

Scheme 4.8 Synthesis of ketone **10** [24]

The relatively low yields and extremely long reaction times (7 days) prompted us to continue to examine alternative reaction conditions for this Heck coupling. Marsais and co-workers have recently reported a Heck reaction using allylpalladium chloride dimer, tri-*o*-tolyphosphine, sodium acetate as the base in toluene and dimethylacetamide (3:1) at 115 °C for 12 h [25]. The authors used these reaction conditions to synthesise potential starting material for azasteroids. Exploiting this, the Heck coupled product **10** was isolated in 82% yield after a relatively short reaction time of 12 h, Scheme 4.9.

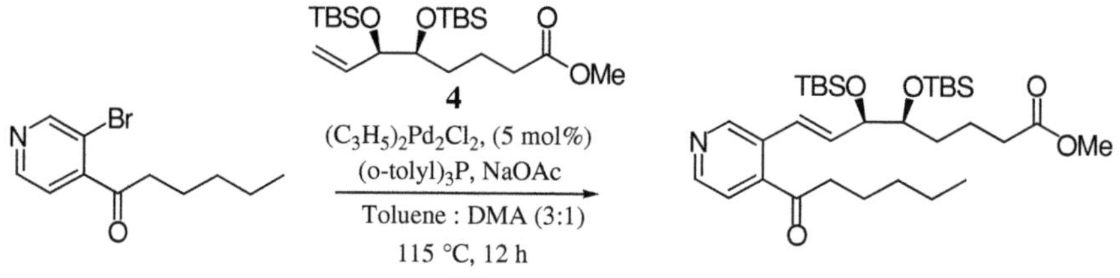

Scheme 4.9 Improved reaction conditions employed used for the synthesis of ketone 10 [13, 25]

The ^{1}H NMR spectrum of ketone **10**, Fig. 4.8, showed the presence of a doublet at 6.75 ppm with a large coupling constant of 16.0 Hz, confirming that the required *E*-stereochemistry had been achieved. The two olefin carbons also showed distinctive peaks in the ^{13}C NMR spectrum at 125.4 and 135.7 ppm. The IR spectrum revealed two carbonyl stretches at 1740 and 1701 cm^{-1}. Attempts to prepare **10** using microwave irradiation were unsuccessful as only starting materials were recovered.

At this point reduction of ketone **10** was required to obtain the alcohol. In the case of the benzene-containing analogues reported by our group, reduction was carried out by using sodium borohydride in MeOH to produce the epimeric alcohols. These were easily separated by column chromatography yielding the

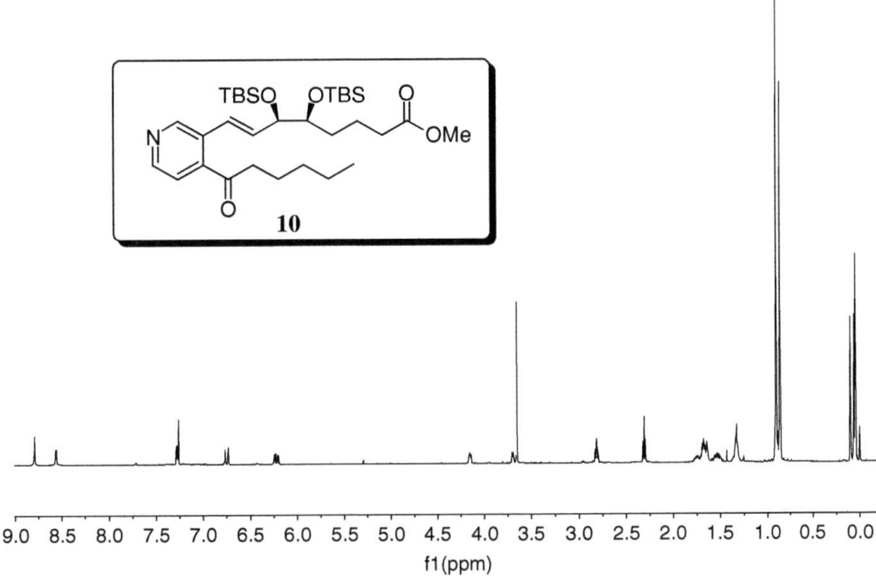

Fig. 4.8   500 MHz ^{1}H NMR spectrum of ketone 10

pure diastereomers. Unfortunately, applying this method of reduction to the pyridine-containing analogues resulted in the formation of two diastereomers which were inseparable by column chromatography. To overcome this problem, ketone **10** was reduced using Brown's (–) and (+) chlorodiisopinocampheylborane [26], affording the desired alcohols (1R)-**14** and (1S)-**14** in 65 and 69% yields, respectively, Scheme 4.10.

Scheme 4.10   Reduction of **10** using (−)-and (+)-DIP-Chloride [26]

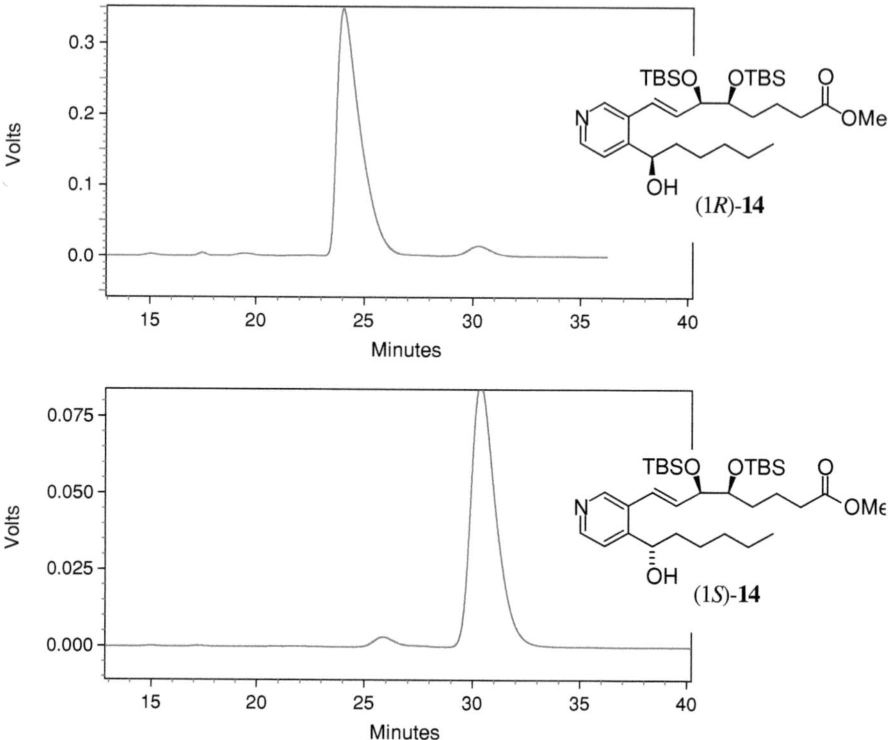

Fig. 4.9 HPLC traces of alcohols (1S)-**14** and (1R)-**14**, performed on a Chiracel® OD column: 99:1 hexane/2-propanol, 1.0 mL/min, t_R = 30.3 min for (1S)-**14**, t_R = 24.7 min for (1R)-14

The formation of the alcohol (1S)-**14** was confirmed by the appearance of a triplet at 4.98 ppm in the ^{1}H NMR spectrum integrating for one proton and corresponding to the CH directly bonded to the hydroxyl group. This CH was also observed in the ^{13}C NMR spectrum at 51.5 ppm. One carbonyl stretch was observed in the IR spectrum at 1742 cm^{-1} along with a broad stretch at 3365 cm^{-1} corresponding to the hydroxyl group. The *de* obtained was 94.9% and 92.3% for alcohols (1S)-**14** and (1R)-**14** respectively, as determined by chiral HPLC, Fig. 4.9.

The final step in the synthesis required the removal of the silyl ether protecting groups. This was achieved under mild conditions, using *p*-toluenesulfonic acid in methanol, providing the pyridine-containing LXA$_4$ (1R)-**2** and (1S)-**2** in 62% and 52% yields, respectively, Scheme 4.11.

The final products (1R)-**2** and (1S)-**2** are extremely polar and therefore require long extraction times from the silica after purification by preparative Thin Layer Chromatography. The formation of the product was confirmed by the absence of the silyl groups in the ^{1}H NMR spectrum, Fig. 4.10. A broad band at 3383 cm^{-1} in the IR spectrum proved the presence of the hydroxyl groups.

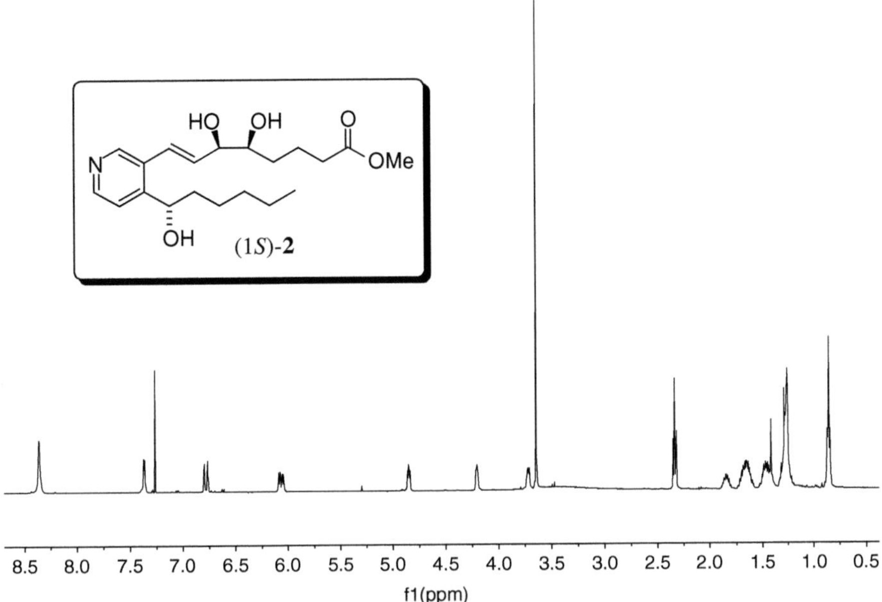

Fig. 4.10 500 MHz ^{1}H NMR spectrum of (1*S*)-**2**

Scheme 4.11 Deprotection to furnish pyridine-containing LXA$_4$ analogues (1*R*)-2 and (1*S*)-2 [13]

(1*S*)-**2**   (1*R*)-**2**

Fig. 4.11 Pyridine-containing LXA$_4$ analogues (1*R*)-2 and (1*S*)-2

4.4 Biological Evaluation of Pyridine-Containing LXA$_4$ Analogues

Having effectively prepared both diastereomers of the pyridine-containing LXA$_4$ analogues, Fig. 4.11, the compounds were evaluated for their ability to promote the clearance of apoptotic PMNs. The results obtained were compared to the native LXA$_4$ and the parent aromatic analogue **1**.

The activity of these compounds were tested by our collaborators from Prof. Catherine Godson's research group in the Conway Institute of Biomolecular and Biomedical Research, University College Dublin. Differentiated THP-1 were exposed to the pyridine LXA$_4$ analogues, (1*R*)-**2** and (1*S*)-**2** at concentrations ranging from 0.1 to 10 nM, for 15 min at 37 °C before addition of apoptotic human PMNs. The extent of phagocytosis was compared with that obtained using native LXA$_4$ (1 nM; 15 min at 37 °C), which was previously shown to significantly enhance phagocytosis, Fig. 4.12.

Pretreatment of differentiated THP-1 cells with compound (1*S*)-**2** at 1 nM and 10 nM resulted in a significant increase of phagocytosis of apoptotic PMNs. These results were comparable to the effect observed with the native LXA$_4$, Fig. 4.13. No effect was observed when a concentration of 0.1 was used, Fig. 4.12a. Compound (1*R*)-**2** significantly stimulated phagocytosis only when used at 1 nM concentration, although there is no statistical difference between the results determined for (1*R*)-**2** and (1*S*)-**2** compared to those obtained with (1*R*)-**1** and (1*S*)-**1**, Fig. 4.12b.

Given that the native LX have previously been reported to affect the production of inflammatory cytokines [27–30] we assessed the ability of our pyridine-containing LXA$_4$ analogues (1*R*)-**2** and (1*S*)-**2** to modulate the production of interleukin-12p40 (IL-12p40), IL-1β and monocyte chemoattractant protein-1 (MCP-1) using a J774 murine macrophage cell line, Fig. 4.13. This research was carried out by our collaborators in Dr. Christine Loscher's research group from the School of Biotechnology, Dublin City University.

Lipopolysaccharide (LPS 100 ng/ml) was used to induce cytokine production in the cells over 24 h. Addition of (1*R*)-**2** at a concentration of 10 μM, 1 h prior to

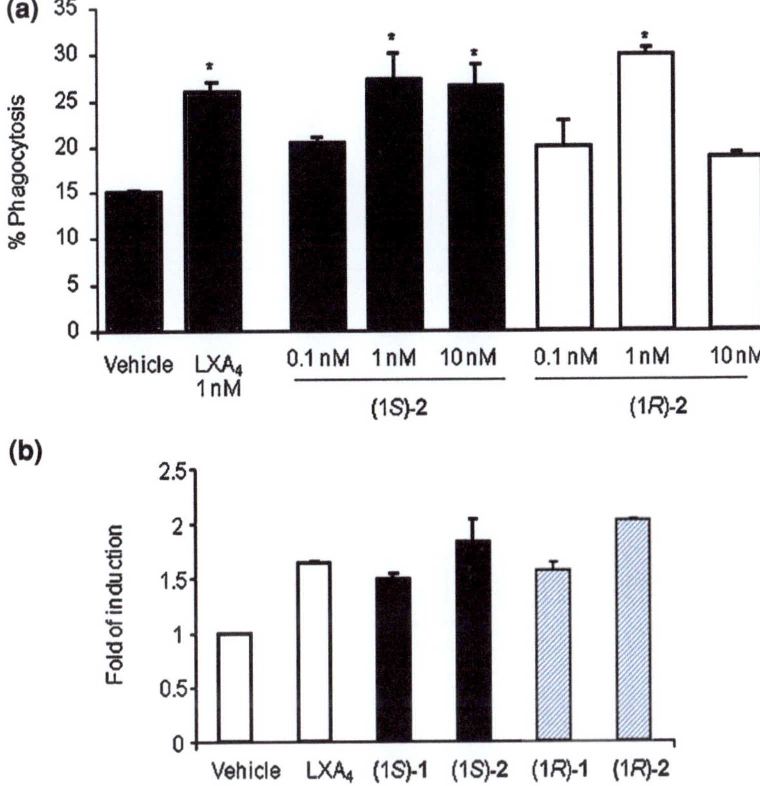

Fig. 4.12   **A**: LXA₄ analogues promote phagocytosis of apoptotic PMNs by differentiated THP-1 cells. Differentiated THP-1 cells (5×10^5) were treated with vehicle (control), LXA₄ (1 nM), or LXA₄ analogues at the concentrations indicated, for 15 min at 37 °C prior to co-incubation with apoptotic PMNs (1×10^6) for 2 h at 37 °C. Phagocytosis was detected by staining of PMN and quantified by light microscopy. Data are expressed as % phagocytosis and represent means ± SEM (n = 3): *p < 0.05 vs vehicle (control).**B**: THP-1 cells (5×10^5) were treated with vehicle (control), LXA₄, benzo analogues or pyridine analogues at the concentration of 1 nM for 15 min at 37 °C prior to co-incubation with apoptotic PMN (1×10^6) for 2 h at 37 °C. Data are expressed as fold of induction over basal and represent means ± SEM (n = 3)

LPS stimulation, resulted in a suppression of IL-12p40, Fig. 4.13. Exposure of cells to (1S)-**2** had a more potent effect on IL–12p40 production, with significant suppression of this cytokine at 10 μM, 1 μM and 1 nM. Both (1R)-**2** and (1S)-**2** suppressed LPS-induced production of IL-1β at both 1 μM and 1 nM concentrations. There was no effect on MCP production, demonstrating that the effects of the analogues were specific.

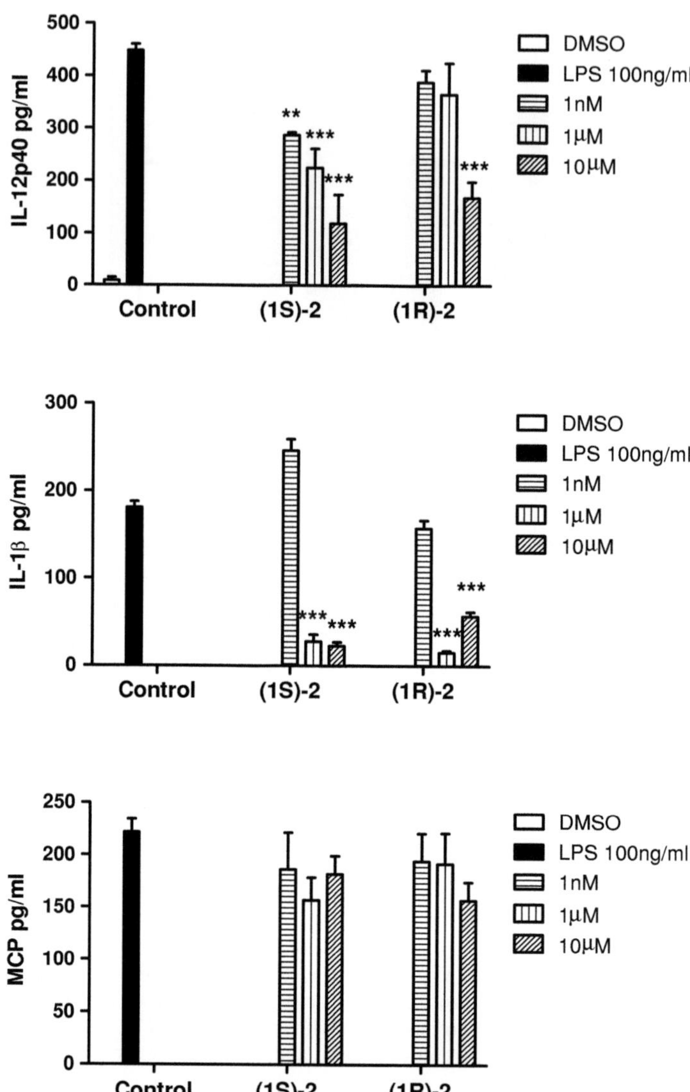

Fig. 4.13 LXA$_4$ analogues suppress pro-inflammatory cytokine production by J774 macrophages. J774 macrophages (1×10^6) were treated with vehicle (control), (1*R*)-2 or (1*S*)-**2** analogues at the concentrations indicated, for 1 h prior to stimulation with LPS (100 ng/ml). The concentrations of cytokines were assessed by ELISA. Data represent means $\pm$ SEM (n = 4): **p < 0.01, ***p < 0.001 determined by one-way ANOVA comparing all groups

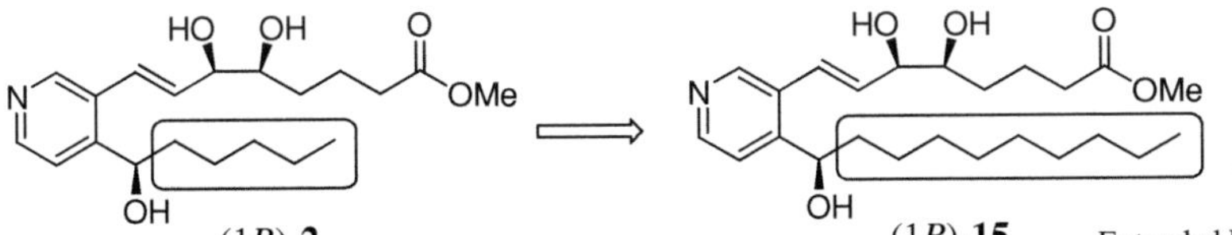

Fig. 4.14   Design of pyridine-containing LXA$_4$ analogue (1*R*)-**15**

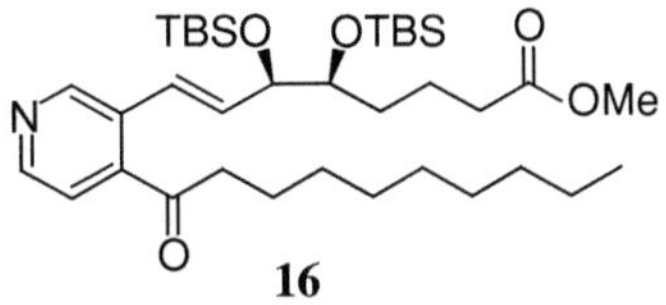

Fig. 4.15   Key intermediate **16**

4.5 Synthesis of Pyridine-Containing LXA$_4$ Analogues with an Extended Lower Chain

Petasis and co-workers. recently reported an increase in bioactivity of benzene-containing LXA$_4$ analogues with an extended lower chain [2, 3], reviewed in Chap. 2. In light of these findings we envisaged using the modular synthetic approach described for our pyridine-containing LXA$_4$ analogues (1*R*)-**2** and (1*S*)-**2**,[13] to synthesise and probe the activity of a pyridine-containing LXA$_4$ analogue (1*R*)-**15** with a longer extended lower chain, Fig. 4.14.

It was foreseen that this analogue would add increased value to our Structure Activity Relationship Study and furthermore, provide some insight into the accepted chain length of the compound in the active site of the receptor. It was hoped that suitable probing of the lower chain would further enhance the biological activity observed with (1*R*)-**2** and (1*S*)-**2**.

The synthetic route devised for the pyridine-containing LXA$_4$ analogue (1*R*)-**15** relied on the formation of ketone **16**, Fig. 4.15.

The key synthetic transformations for the construction of ketone **16** included, as before, a regiospecfic pyridine lithiation and a palladium-catalysed Heck reaction, Scheme 4.12.

The formation of the 3,4-disubstituted pyridine **17** was obtained by the lithiation procedure reported by Gribble and Saulnier [14], followed by quenching with decanal. Poorer yields, 31%, were obtained compared to the quenching with hexanal, 75%. This isolated alcohol was further oxidised with PCC in the presence of glacial acetic acid and the product was determined by the appearance of a triplet in the ^{1}H NMR spectrum, Fig. 4.16, at 2.87 ppm integrating for the two protons corresponding to the CH$_2$ directly attached to the carbonyl. This was also accompanied with a peak at 202.6 ppm in the ^{13}C NMR spectrum and a

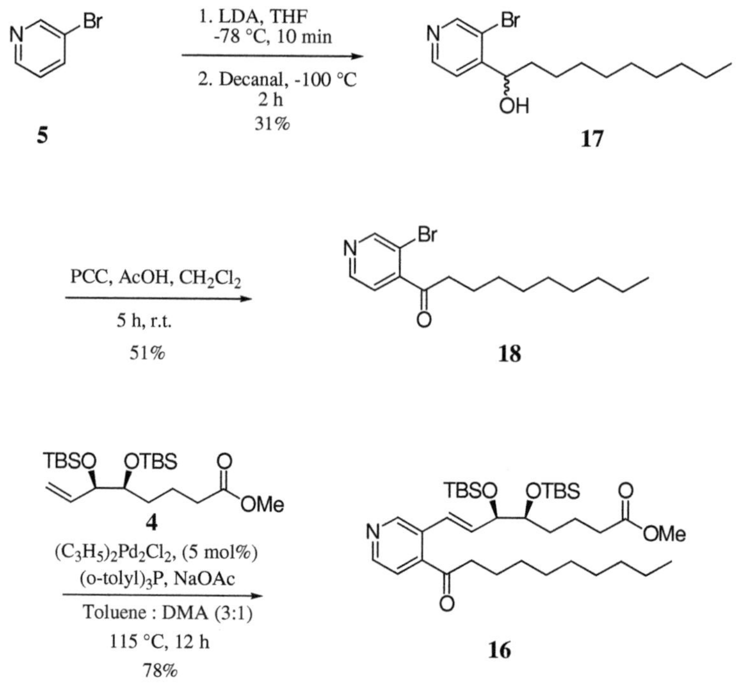

Scheme 4.12 Synthesis of **16**

characteristic sharp stretch at 1709 cm^{-1} in the IR spectrum confirming the formation of ketone **18**.

The Heck coupled product **16** was isolated in 78% yield following use of the reaction conditions reported by Marsais [25]. The presence of a doublet at 6.75 ppm with a large coupling constant of 16.1 Hz in the ^{1}H NMR spectrum confirmed the required *E*-stereochemistry was in place. The two olefin carbons were apparent in the ^{13}C NMR spectrum at 125.5 and 135.7 ppm. The IR spectrum revealed two carbonyl stretches at 1739 and 1698 cm^{-1}.

Ketone **16** was reduced using sodium borohydride in MeOH to give epimeric alcohols (1*R/S*)-**19**. An asymmetric reduction was also carried out using Brown's (+) chlorodiisopinocampheylborane [26] to furnish alcohol (1*R*)-**19**, Scheme 4.13.

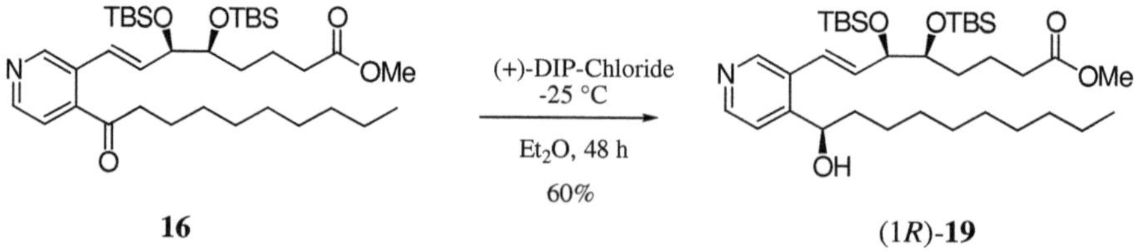

Scheme 4.13 Preparation of alcohol (1*R*)-**19**

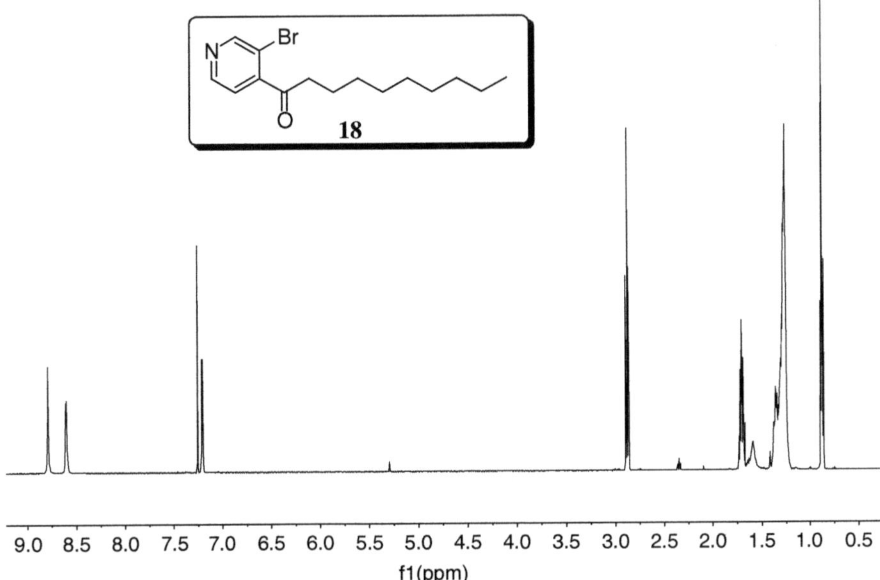

Fig. 4.16   500 MHz ^{1}H NMR spectrum of ketone **18**

The ^{1}H NMR spectrum of the alcohol (1R/S)-**19** was used in order to determine the *de* value of 93%, Fig. 4.17. Specifically, the integration of the doublets at 6.75 and 6.65 ppm were used to determine this *de* value. Further confirmation of this *de* value was determined by chiral HPLC.

Finally, alcohol (1R)-**19** was deprotected under the mild conditions of *p*-toluenesulfonic acid in methanol affording pyridine LXA$_4$ (1R)-**15** in 51% yield, Scheme 4.14.

Scheme 4.14   Removal of the silyl ether protection groups

The formation of the product was verified by the disappearance of the silyl protecting group protons and carbons in the ^{1}H and ^{13}C NMR spectra. This analogue is currently being evaluated for its ability to promote phagocytosis of apoptotic PMNs along with its ability to suppress key pro-inflammatory cytokines. The pending biological results will reveal if the extended lower chain has a positive effect on the bioactivity compared to the (1R)-**2** analogue.

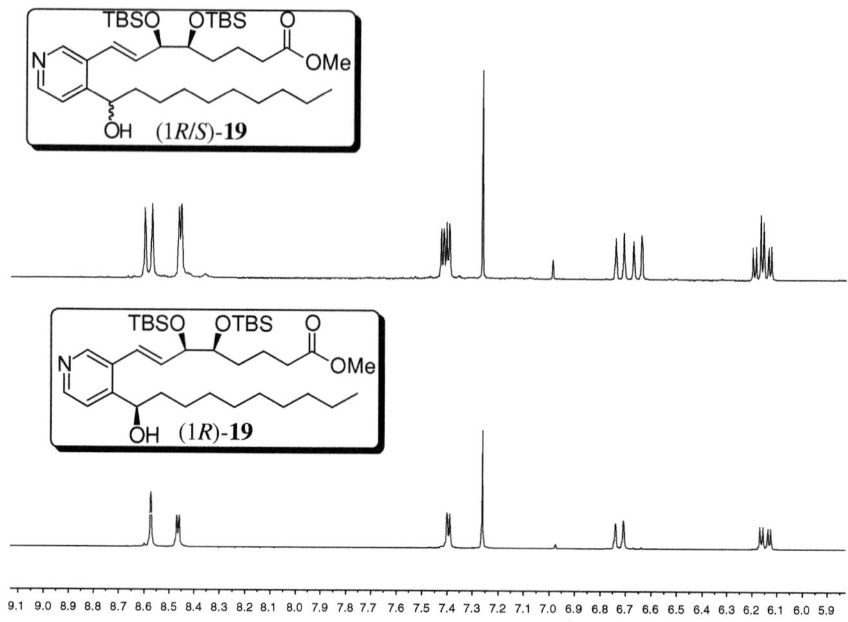

Fig. 4.17 500 MHz ^{1}H NMR spectrum of alcohol **19**

4.6 Conclusion

In summary, we have described the synthesis of a novel class of Lipoxin A_4 analogues where the unstable triene system has been replaced by a pyridine ring. The pyridine-containing Lipoxin analogues induced a greater increase in phagocytosis of PMNs by macrophages compared to both the natural Lipoxin A_4 and the benzene-containing Lipoxin analogue. Furthermore, they displayed anti-inflammatory characteristics, demonstrated by their suppression of pro-inflammatory cytokine production by macrophages. We have successfully used our modular synthetic approach to produce a novel pyridine-containing Lipoxin A_4 analogue with an extended lower chain of 10 carbons instead of the conventional 6 carbon chain which is native to the naturally occurring LX. Exploiting this analogue, we hope to add increased value to our Structure Activity Relationship Study and furthermore provide some insight into the acceptable chain length of the compound in the active site of the receptor.

4.7 Experimental

4.7.1 1-(3-Bromopyridin-4-yl)hexan-1-ol (7)

n-Butylithium (2.8 mL, 2.5 M in hexanes, 6.9 mmol) was added to a solution of diisopropylamine (0.88 ml, 6.3 mmol) in THF (20 ml) at $-78\ ^\circ$C under an atmosphere of nitrogen and stirring was continued for 15 min. 3-Bromopyridine **5** (0.62 ml, 6.3 mmol) in THF (1 ml) was added over 10 min (maintaining the internal temperature below $-75\ ^\circ$C). The reaction was brought to $-100\ ^\circ$C for 10 min and hexanal **6** (1.26 g, 12.6 mmol) in THF (3 ml) was added over 10 min (again maintaining the internal temperature below $-75\ ^\circ$C). The reaction mixture was stirred at $-100\ ^\circ$C for 1 h and then warmed to $-20\ ^\circ$C over 20 min. The mixture was quenched with a saturated ammonium chloride solution (3 ml) and extracted using diethyl ether (3 × 25 ml), washed with water (25 ml), brine (25 ml) and dried over sodium sulfate. The solvent was removed in vacuo and the residue was purified using silica gel chromatography (pentane/ethyl acetate, 9:1 then 4:1) to afford **7** (737 mg, 75% yield) as a viscous yellow oil. TLC: $R_f = 0.21$ (pentane/ethyl acetate, 4:1); ^{1}H NMR (500 MHz, CDCl$_3$) ppm 8.57 (s, 1H), 8.43 (d, $J = 4.9$ Hz, 1H), 7.51(d, $J = 4.9$ Hz, 1H), 4.98 (m, 1H), 3.11 (br. s, 1H, exchanges with D$_2$O), 1.79-1.73 (m, 1H), 1.65–1.49 (m, 1H), 1.49–1.40 (m, 2H), 1.39- 1.26 (m, 4H), 0.9 (t, $J = 7.0$ Hz, 3H); ^{13}C NMR (125 MHz, CDCl$_3$) ppm 153.3, 151.5, 148.3, 122.2, 120.0, 71.9, 37.1, 31.5, 25.3, 22.5, 14.0; IR (neat) (v_{max}, cm^{-1}) 3583, 3296, 2929, 2361, 1588, 1466, 1401, 1343, 1217, 1162, 1084, 756; HRMS (EIMS) Found 258.0486 [M + H]$^+$ C$_{11}$H$_{17}$BrNO requires 258.0494.

4.7.2 1-(3-Bromopyridin-4-yl)hexan-1-one (3)

Glacial acetic acid (0.21 ml) was added to a vigorously stirred solution of pyridinium chlorochromate (821 mg, 3.81 mmol) in dry dichloromethane (20 ml).

After 5 min at room temperature, alcohol **7** (655 mg, 2.54 mmol) in dichloromethane (5 ml) was added and the mixture was stirred at room temperature for 5 h. Diethyl ether (40 ml) was added and the mixture was gravity filtered twice with filter paper. The solvent was removed in vacuo and the residue was purified using silica gel chromatography (pentane/ethyl acetate, 4:1) to afford ketone **3** (458 mg, 70% yield) as an orange oil. TLC: $R_f = 0.67$ (pentane/ethyl acetate, 4:1); ^{1}H NMR (500 MHz, CDCl$_3$) ppm 8.79 (s, 1H), 8.60 (d, $J = 4.9$ Hz, 1H), 7.22 (d, $J = 4.9$ Hz, 1H), 2.88 (t, $J = 7.4$ Hz, 2H), 1.74–1.69 (m, 2H), 1.37–1.33 (m, 4H), 0.91 (t, $J = 7.1$ Hz, 3H); ^{13}C NMR (125 MHz, CDCl$_3$) ppm 202.6, 152.8, 148.5, 148.5, 121.6, 116.0, 42.5, 31.2, 23.2, 22.3, 13.8; IR (neat) (v_{max}, cm^{-1}) 2958, 2931, 1710, 1578, 1466, 1396, 1378, 1276, 1250, 1089, 1022; HRMS (EIMS) Found 256.0336 [M + H]$^+$ C$_{11}$H$_{15}$BrNO requires 256.0337.

4.7.3 (5S, 6R, E)-Methyl 5,6-bis(tert-butyldimethylsilyloxy)-8-(4-hexanoylpyridin-3-yl)oct-7-enoate (10)

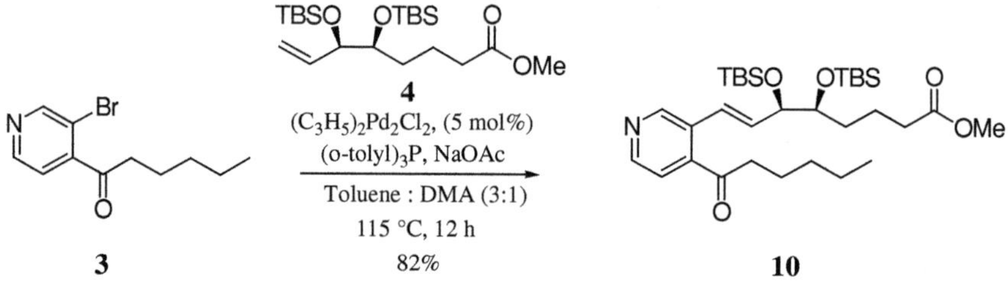

$[\eta^3$- (C$_3$H$_4$)Pd(μ-Cl)$_2$]$_2$ (17 mg, 0.048 mmol), P(o-tolyl)$_3$ (34 mg, 0.096 mmol) and NaOAc (234 mg, 2.88 mmol) were dissolved in dry freshly distilled toluene (2 ml) to which ketone **3** (250 mg, 0.96 mmol) in toluene (1 ml) and olefin **4** (406 mg, 0.96 mmol) in toluene (1 ml) were added. DMA (1.3 ml) was added and the reaction mixture was sealed under nitrogen and stirring was continued for 12 h at 115 °C followed by filtration through a pad of Celite$^®$. The solvent was removed in vacuo and the residue was purified using silica gel chromatography (pentane, then pentane/diethyl ether, 9:1 then 4:1, then 3:2) to afford **10** (486 mg, 82% yield) as a viscous yellow oil. TLC: $R_f = 0.23$ (pentane/diethyl ether, 3:2); $[\alpha]_D^{20}$ -12.9 ($c = 0.84$, CH$_2$Cl$_2$); ^{1}H NMR (500 MHz, CDCl$_3$) ppm 8.79 (s, 1H), 8.56 (d, $J = 5.0$ Hz, 1H), 7.27 (d, $J = 5.0$ Hz, 1H), 6.75 (d, $J = 16.0$ Hz, 1H), 6.22 (dd, $J = 16.0$, 6.5 Hz, 1H), 4.16 (dd, $J = 6.5$, 4.8 Hz, 1H), 3.70 (m, 1H), 3.66 (s, 3H), 2.81 (dt, $J = 7.3$, 9.5 Hz, 2H), 2.31 (t, $J = 7.4$, 2H), 1.31–1.79 (m, 10H), 0.91 (s, 9H), 0.87 (s, 12H), 0.05–0.10 (3 x s, 12H); ^{13}C NMR (125 MHz, CDCl$_3$) ppm 204.1, 173.9, 149.0, 148.4, 144.2, 135.7, 129.9, 125.4, 120.4, 77.0, 76.2, 51.5, 42.4, 34.2, 33.0, 31.3, 26.0, 23.6, 22.5, 20.7, 18.3, 18.2, 13.9, −4.0, −4.2, −4.6, −4.7; IR (neat) (v_{max}, cm^{-1}) 2995, 2929, 2857, 1740, 1701; HRMS (EIMS) Found 592.3870 [M + H]$^+$ C$_{32}$H$_{58}$NO$_5$Si$_2$ requires 592.3881.

4.7.4 (5S, 6R, E)-Methyl 5,6-bis(tert-butyldimethylsilyloxy) -8-(4-((R)-1-hydroxyhexyl)pyridin-3-yl)oct-7-enoate ((1S)-14)

Ketone **10** (97 mg, 0.163 mmol) in diethyl ether (1 ml) was added to a solution of (−) DIPCl (210 mg, 0.64 mmol) in diethyl ether (1 ml) at -25 °C, and stirring was continued for 48 h. The reaction mixture was diluted with pentane (1 ml) and diethyl ether (1 ml) and diethanol amine (34 mg, 0.326 mmol) was added. Stirring was continued for 4 h at room temperature, followed by filtration and removal of the solvent in vacuo. The residue was purified by silica gel chromatography (pentane, then pentane/diethyl ether, 1:1, then diethyl ether/pentane, 2:1) to afford (1S)-**14** (67 mg, 69% yield) as a colourless oil. de = 94.9 as determined by chiral HPLC using an OD column (Hexane : iPrOH, 99:1) flow rate : 1 ml/min, 24.0 min for (R), 30.3 min for (S); TLC: R_f = 0.15 (diethyl ether/pentane, 2:1); $[\alpha]_D^{20}$ -31.9 (c = 0.56, CHCl$_3$); ^{1}H NMR (500 MHz, CDCl$_3$) ppm 8.60 (s, 1H), 8.45 (d, J = 5.0 Hz, 1H), 7.41 (d, J = 5.0 Hz, 1H), 6.65 (d, J = 15.8 Hz, 1H), 6.18 (dd, J = 15.8, 6.5 Hz, 1H), 4.98 (t, J = 6.2 Hz, 1H) 4.18 (dd, J = 6.5, 4.9 Hz, 1H), 3.70 (m, 1H), 3.65 (s, 3H), 2.31 (t, J = 7.3, 2H), 1.27–1.80 (m, 12H), 0.92 (s, 9H), 0.87 (s, 12H), 0.10–0.05 (3 x s, 12H); ^{13}C NMR (125 MHz, CDCl$_3$) ppm 174.0, 150.3, 148.7, 147.6, 134.8, 130.5, 124.5, 119.8, 77.2, 76.2, 69.8, 51.5, 38.1, 34.3, 33.0, 31.7, 26.0, 26.0, 25.4, 22.6, 20.7, 18.3, 18.2, 14.0, −4.0, −4.1, −4.6, −4.6; IR (neat) (v_{max}, cm^{-1}) 3350, 2954, 2930, 2857, 1742, 1252; HRMS (EIMS) Found 594.3992 [M + H]$^+$ C$_{32}$H$_{60}$NO$_5$Si$_2$ requires 594.4010.

4.7.5 (5S, 6R, E)-Methyl 5,6-bis(tert-butyldimethylsilyloxy) -8-(4-((R)-1-hydroxyhexyl)pyridin-3-yl)oct-7-enoate ((1R)-14)

Ketone **10** (190 mg, 0.32 mmol) in diethyl ether (2 ml) was added to a solution of (+) DIPCl (0.41 g, 1.28 mmol) in diethyl ether (2 ml) at -25 °C, and stirring was continued for 48 h. The reaction mixture was diluted with pentane (2 ml) and diethyl ether (2 ml) and diethanol amine (64.3 mg, 0.64 mmol) was added and stirring was continued for 4 h at room temperature followed, by filtration and removal of the solvent in vacuo. The residue was purified by silica gel chromatography (pentane, then pentane/diethyl ether, 1:1 then diethyl ether/pentane, 2:1) to afford (1R)-**14** (125 mg, 65% yield) as a viscous yellow oil. *de* = 92.3%, as determined by chiral HPLC using an OD column (hexane: *i*PrOH, 99:1) flow rate: 1 ml/min, 24.0 min for (*R*), 30.3 min for (*S*). TLC: R_f = 0.15 (diethyl ether/ pentane, 2:1); $[\alpha]_D^{20}$ +29.4 (*c* = 0.36, CHCl$_3$); ^{1}H NMR (500 MHz, CDCl$_3$) ppm 8.57 (s, 1H), 8.45 (d, *J* = 5.0 Hz, 1H), 7.40 (d, *J* = 5.0 Hz, 1H), 6.72 (d, *J* = 15.8 Hz, 1H), 6.15 (dd, *J* = 15.8, 6.0 Hz, 1H), 4.94 (t, *J* = 6.7 Hz, 1H) 4.22 (m, 1H), 3.71 (m, 1H), 3.64 (s, 3H), 2.30 (t, *J* = 7.5 Hz, 2H), 1.27–1.79 (m, 12H), 0.93 (s, 9H), 0.88 (s, 12H), 0.11–0.06 (3 x s, 12H); ^{13}C NMR (125 MHz, CDCl$_3$) ppm 174.0, 151.0, 147.9, 147.0, 135.2, 130.7, 124.3, 120.2, 77.0, 76.2, 70.2, 51.5, 37.9, 34.2, 32.8, 31.7, 25.9, 25.4, 22.6, 20.9, 18.3, 18.2, 13.9, −4.0, −4.3, −4.6, −4.7; IR (neat) (ν_{max}, cm^{-1}) 3365, 2954, 2930, 2857, 1741, 1252; HRMS (EIMS) Found 594.4033 [M + H]$^+$ C$_{32}$H$_{60}$NO$_5$Si$_2$ requires 594.4010.

4.7.6 (5S, 6R, E)-Methyl 5,6-dihydroxy-8-(4-((S) -1-hydroxyhexyl)pyridin-3-yl)oct-7-enoate ((1S)-2)

TBSO   OTBS   O
N
OMe
OH
(1S)-14
p-TSA, MeOH,
72h, 30 °C
52 %
HO   OH   O
N
OMe
OH
(1S)-2

Alcohol (1S)-**14** (50 mg, 0.084 mmol) was dissolved in dry MeOH (1 ml) to which *p*-toluenesulfonic acid (24.1 mg, 0.126 mmol) and the mixture was stirred at 30 °C for 48 h. The solvent was removed in vacuo at 30 °C to prevent formation of a lactone by-product and the residue was purified by silica gel chromatography (pentane/ethyl acetate, 1:1,then ethyl acetate, then ethyl acetate/MeOH, 99:1) to afford (1S)-**2** (16.4 mg, 52% yield) as a colourless viscous oil. TLC: R_f = 0.26 (CH$_2$Cl$_2$/MeOH, 9.5:0.5); $[\alpha]_D^{20}$ -8.8 (*c* = 0.34, CHCl$_3$); ^{1}H NMR (500 MHz, CDCl$_3$) ppm 8.36 (br. s, 2H), 7.36 (d, *J* = 5.0 Hz, 1H), 6.77 (d, *J* = 15.9 Hz, 1H), 6.07 (dd, *J* = 15.9, 6.4 Hz, 1H), 4.85 (dd, *J* = 7.3, 5.2 Hz, 1H) 4.21 (m, 1H), 3.72 (m, 1H), 3.65 (s, 3H), 2.34 (t, *J* = 7.3, 2H), 1.26–1.89 (m, 13H), 0.86 (t, *J* = 6.7 Hz, 3H); ^{13}C NMR (125 MHz, CDCl$_3$) ppm 174.3, 151.3, 148.2, 147.2, 132.8, 130.7, 126.1, 120.6, 75.4, 73.9, 70.1, 51.6, 37.6, 33.7, 31.7, 31.6, 25.4, 22.5,

21.1, 14.0.IR (neat) (v_{max}, cm^{-1}) 3383, 2954, 2857, 1733, 1460, 1259; HRMS (EIMS) Found 366.2291 [M + H]$^+$ C$_{20}$H$_{32}$NO$_5$ requires 366.2280.

4.7.7 (5S, 6R, E)-Methyl 5,6-dihydroxy-8-(4-((R)-1-hydroxyhexyl)pyridin-3-yl)oct-7-enoate ((1R)-2)

Alcohol (1R)-**14** (160 mg, 0.266 mmol) was dissolved in dry MeOH (1 ml) to which p–toluenesulfonic acid (77 mg, 0.4 mmol) was added and the mixture was stirred at 30 °C for 48 h. The solvent was removed in vacuo at 30 °C to prevent formation of a lactone by-product and the residue was purified by silica gel chromatography (pentane/ethyl acetate, 1:1, then ethyl acetate, then ethyl acetate/ MeOH) 99:1 to afford (1R)-**2** (60.8 mg, 62% yield) as a colourless viscous oil. TLC: R$_f$ = 0.26 (CH$_2$Cl$_2$/MeOH, 9.5:0.5); [α]$_D^{20}$ +12.2 (c = 2.3, CHCl$_3$); ^{1}H NMR (500 MHz, CDCl$_3$) ppm 8.39 (s, 1H), 8.35 (ap d, J = 4.8 Hz 1H) 7.38 (d, J = 4.8 Hz, 1H), 6.76 (d, J = 15.8 Hz, 1H), 6.15 (dd, J = 15.8, 6.1 Hz, 1H), 4.88 (br. s, 1H) 4.22 (br. s, 1H), 3.76 (br. s, 1H), 3.65 (s, 2H), 3.01 (br. s, 3H), 2.35 (t, J = 7.27, 2H), 1.26–1.89 (m, 13H), 0.86 (t, J = 6.58 Hz, 3H); ^{13}C NMR (125 MHz, CDCl$_3$) 174.3, 152.2, 147.4, 146.4, 132.9, 130.8, 125.9, 120.7, 75.6, 74.0, 69.3, 51.6, 37.5, 33.7, 31.9, 31.6, 25.3, 22.5, 21.2, 14.0. IR (neat) (v_{max}, cm^{-1}) 3389, 2953, 2928, 2856, 1737, 1457, 1235; HRMS (EIMS) Found 366.2278 [M + H]$^+$ C$_{20}$H$_{32}$NO$_5$ requires 366.2280.

4.7.8 1-(3-Bromopyridin-4-yl)decan-1-ol (17)

n-Butyllithium (c = 2.5 M, 5.6 ml, 13.9 mmol) was added to a solution of diiso-propylamine (1.78 mL, 12.6 mmol) in THF (40 ml) at −78 °C under an atmosphere of nitrogen and stirring was continued for 15 min. 3-Bromopyridine **5** (1.2 ml, 12.6 mmol) in THF (6 ml) was added over 10 min (maintaining the internal temperature below -75 °C). The reaction was brought to -100 °C for 10 min and decanal (4.73 ml, 25.2 mmol) in THF (5 ml) was added over 10 min (again maintaining the internal temperature below -75 °C). The reaction mixture was stirred at -100 °C for 1 h and then warmed to -20 °C over 20 min. The mixture was quenched with a saturated ammonium chloride solution (6 ml) and extracted using diethyl ether (3 × 50 ml), washed with water (50 ml), brine (50 ml) and dried over sodium sulfate. The solvent was removed in vacuo and the residue was purified using silica gel chromatography (pentane/ethyl acetate, 9:1 then 4:1) to afford **17** as a viscous yellow oil. (1.21 g, 31% yield) as a viscous colourless oil. TLC: $R_f = 0.25$ (pentane/ethyl acetate, 4:1) ^{1}H NMR (500 MHz, CDCl$_3$) ppm 8.59 (s, 1H), 8.45 (d, $J = 5.0$ Hz, 1H), 7.51(d, $J = 5.0$ Hz, 1H), 4.98 (m, 1H), 2.79 (br s, 1H,), 1.80–1.73 (m, 1H), 1.64–1.57 (m, 1H), 1.51–1.26 (m, 14H), 0.87 (t, $J = 6.8$ Hz, 3H); ^{13}C NMR (125 MHz, CDCl$_3$) ppm 153.4, 151.4, 148.3, 122.2, 120.0, 72.0, 37.1, 31.9, 29.52, 29.50, 29.3, 29.2, 25.6, 22.7, 14.1; IR (neat) (ν_{max}, cm^{-1}) 3258, 2924, 2855, 1586, 1463, 1401, 1080; HRMS (EIMS) Found 314.1129 [M + H]$^+$ C$_{15}$H$_{25}$BrNO requires 314.1120.

4.7.9 1-(3-Bromopyridin-4-yl)decan-1-one (18)

Glacial acetic acid (0.4 ml) was added to a vigorously stirred solution of pyridinium chlorochromate (688 mg, 1.59 mmol) in dry dichloromethane (10 ml). After 5 min at room temperature, alcohol **17** (500 mg, 1.59 mmol) in dichloromethane (5 ml) was added and the mixture was stirred at room temperature for 5 h. Diethyl ether (40 ml) was added and the mixture was gravity filtered twice with filter paper. The solvent was removed in vacuo and the residue was purified using silica gel chromatography (pentane/ethyl acetate, 4:1) to afford ketone **18** (246 mg, 49% yield) as viscous colourless oil. TLC: $R_f = 0.61$ (pentane/ethyl acetate, 4:1) ^{1}H NMR (500 MHz, CDCl$_3$) ppm 8.78 (s, 1H), 8.59 (d, $J = 4.8$ Hz, 1H), 7.21 (d, $J = 4.8$ Hz, 1H), 2.87 (t, $J = 7.4$ Hz, 2H), 1.72–1.67 (m, 2H), 1.37–1.25 (m, 12H), 0.87 (t, $J = 6.7$ Hz, 3H); ^{13}C NMR (125 MHz, CDCl$_3$) ppm 202.6, 152.8, 148.6, 148.5, 121.6, 116.1, 42.6, 31.8, 29.4, 29.3, 29.2, 29.1, 23.6, 22.6, 14.1; IR (neat) (ν_{max}, cm^{-1}) 2924, 2854, 1709, 1466, 1395, 1226, 1169; HRMS (EIMS) Found 311.0880 [M] C$_{15}$H$_{22}$BrNO requires 311.0885.

4.7.10 (5S, 6R, E)-Methyl 5,6-bis(tert-butyldimethylsilyloxy)-8-(4-decanoylpyridin-3-yl)oct-7-enoate (16)

$[\eta^3\text{-}(C_3H_4)Pd(\mu\text{-}Cl)_2]_2$ (7 mg, 0.019 mmol), $P(o\text{-tolyl})_3$ (14 mg, 0.035 mmol) and NaOAc (95 mg, 1.17 mmol) were dissolved in dry freshly distilled toluene (1 ml) to which ketone **18** (182 mg, 0.58 mmol) in toluene (1 ml) and olefin **4** (1.62 mg, 0.39 mmol) in toluene (1 ml) were added. DMA (1 ml) was added and the reaction mixture was sealed under nitrogen and stirring was continued for 12 h at 115 °C followed by filtration through a pad of Celite$^{®}$. The solvent was removed in vacuo and the residue was purified using silica gel chromatography (pentane, then pentane/diethyl ether, 9:1 then 4:1, then 3:2) to afford **16** (198 mg, 78% yield) as a viscous yellow oil. TLC: $R_f = 0.41$ (pentane/diethyl ether, 3:2); $[\alpha]_D^{20}$ -14.4 ($c = 0.92$, $CHCl_3$); ^1H NMR (500 MHz, $CDCl_3$) ppm 8.80 (s, 1H), 8.56 (d, $J = 4.1$ Hz, 1H), 7.27 (d, $J = 4.1$ Hz, 1H), 6.75 (d, $J = 16.1$ Hz, 1H), 6.22 (dd, $J = 16.1$, 6.7 Hz, 1H), 4.15 (m, 1H), 3.70 (m, 1H), 3.66 (s, 3H), 2.81 (dt, $J = 7.2$, 2.0 Hz, 2H), 2.31 (t, $J = 7.4$, 2H), 1.75- 1.26 (m, 18H), 0.91 (s, 9H), 0.87 (s, 12H), 0.10-0.05 (3 x s, 12H); ^{13}C NMR (125 MHz, $CDCl_3$) ppm 204.1, 173.9, 149.1, 148.5, 144.2, 135.7, 129.9, 125.5, 120.3, 77.1, 76.3, 51.5, 42.5, 33.1, 31.9, 29.4, 29.3, 29.2, 26.0, 26.0, 24.0, 22.7, 20.8, 18.3, 18.2, 14.1, -4.0, -4.1, -4.6, -4.6.; IR (neat) (v_{max}, cm^{-1}) 2927, 2855, 1739, 1698, 1463, 1364, 1251, 1162; HRMS (EIMS) Found 648.4477 $[M + H]^+$ $C_{36}H_{66}NO_5Si_2$ requires 648.4480.

4.7.11 (5S, 6R, E)-Methyl 5,6-bis(tert-butyldimethylsilyloxy)-8-(4-((R)-1-hydroxydecyl)pyridin-3-yl)oct-7-enoate ((1R)-19)

Ketone **16** (150 mg, 0.23 mmol) in diethyl ether (1 ml) was added to a solution of (+) DIPCl (297 mg, 0.93 mmol) in diethyl ether (2 ml) at -25 °C, and stirring was continued for 48 h. The reaction mixture was diluted with pentane (2 ml) and diethyl ether (2 ml) and diethanol amine (73 mg, 0.65 mmol) was added and stirring was continued for 4 h at room temperature followed, by filtration and removal of the solvent in vacuo. The residue was purified by silica gel chromatography (pentane, then pentane/diethyl ether, 1:1 then diethyl ether/pentane, 2:1) to afford (1*R*)-**19** as a viscous yellow oil (90 mg, 60% yield) *de* = 93%, as determined by chiral HPLC using an OD column (hexane: *i*PrOH, 98:2) flow rate: 1 ml/min, 13.7 min for (*R*), 16.7 min for (*S*). TLC: R_f = 0.48 (diethyl ether/pentane, 1:1); $[\alpha]_D^{20}$ +13.4 (*c* = 0.49, CHCl$_3$); ^{1}H NMR (500 MHz, CDCl$_3$) ppm 8.57 (s, 1H), 8.46 (d, *J* = 5.0 Hz, 1H), 7.39 (d, *J* = 5.0 Hz, 1H), 6.71 (d, *J* = 15.8 Hz, 1H), 6.15 (dd, *J* = 15.8, 6.0 Hz, 1H), 4.92 (t, *J* = 6.1 Hz, 1H), 4.22 (m, 1H), 3.71 (m, 1H), 3.64 (s, 3H), 2.30 (t, *J* = 7.5 Hz, 2H), 2.21 (br. s, 1H,) 1.75- 1.25 (m, 20H), 0.93 (s, 9H), 0.88 (s, 12H),0.11-0.06 (3 x s, 12H); ^{13}C NMR (125 MHz, CDCl$_3$) ppm 174.0, 150.3, 148.7, 147.7, 134.9, 130.5, 124.6, 119.9, 77.1, 76.2, 70.1, 51.5, 38.1, 34.2, 32.9, 31.9, 29.61, 29.56, 29.3, 26.0, 25.8, 22.7, 21.0, 18.3, 18.2, 14.1, -3.9, -4.2, -4.5, -4.6; IR (neat) (v_{max}, cm^{-1}) 3408, 2926, 2855, 1740, 1464,1252; HRMS (EIMS) Found 650.4648 [M + H]$^+$ C$_{36}$H$_{68}$NO$_5$Si$_2$ requires 650.4636.

4.7.12 (5S, 6R, E)-Methyl 5,6-dihydroxy-8-(4-((R) -1-hydroxydecyl)pyridin-3-yl)oct-7-enoate ((1R)-15)

Alcohol (1*R*)-**19** (77 mg, 0.119 mmol) was dissolved in dry MeOH (1 ml) to which *p*–toluenesulfonic acid (47 mg, 0.179 mmol) was added and the mixture was stirred at 30 °C for 48 h. The solvent was removed in vacuo at 30 °C to prevent formation of a lactone by-product and the residue was purified by silica gel chromatography (pentane/ethyl acetate, 1:1, then ethyl acetate, then ethyl acetate/MeOH) 98:2 to afford (1*R*)-**15** (23 mg, 51% yield) as a colourless viscous oil. TLC: R_f = 0.64 (CH$_2$Cl$_2$/MeOH, 9:1); $[\alpha]_D^{20}$ +26.4 (*c* = 0.65, CHCl$_3$); ^{1}H NMR (600 MHz, CDCl$_3$) ppm 8.34 (s, 1H), 8.32 (d, *J* = 5.2 Hz 1H) 7.36 (d, *J* = 5.2 Hz, 1H), 6.74 (d, *J* = 15.8 Hz, 1H), 6.12 (dd, *J* = 15.8, 6.8 Hz, 1H), 4.85(dd, *J* = 7.8, 5.0 Hz, 1H) 4.19 (br. s, 1H), 3.75 (m, 1H), 3.65 (s, 3H), 2.34 (t, *J* = 7.4, 2H), 1.89-1.24 (m, 20H), 0.87 (t, *J* = 6.9 Hz, 3H); ^{13}C NMR (125 MHz, CDCl$_3$) 174.2, 151.3, 148.2, 147.1, 132.4, 130.5, 126.3, 120.5, 75.6, 74.0, 69.6, 51.6, 37.7, 33.7, 31.9, 31.8, 29.6, 29.6, 29.6, 29.3, 25.7, 22.7, 21.2, 14.1; IR (neat)

(ν_{max}, cm^{-1}) 3304, 2924, 2853, 1741, 1435, 1170; HRMS (EIMS) Found 422.2886 [M + H]$^+$ C$_{24}$H$_{40}$NO$_5$ requires 422.2906.

4.7.13 Phagocytosis of Apoptotic PMNs by THP-1 Cells

The human myelomonocytic cell line THP-1 (European Collection of Cell Cultures, Salisbury, UK) was maintained as a suspension of RPMI 1640 supplemented with 2 mmol/L glutamine, 100 IU/ml penicillin, 100 µg/ml streptomycin, and 10% fetal calf serum (Life Technologies Inc, Grand Island, NY). THP-1 cells at 5×10^5/mL were differentiated to a macrophage-like phenotype by treatment with 100 nM phorbol 12-myristate, 13-acetate (PMA) for 48 h at 37 °C. Human PMNs were isolated from peripheral venous blood drawn from healthy volunteers, after informed written consent. Briefly, PMNs were separated by centrifugation on Ficoll-Paque (Pharmacia, Uppsala, Sweden) followed by dextran sedimentation (Dextran T500; Pharmacia) and hypotonic lysis of red cells. PMNs were suspended at 4×10^6 cells/mL and spontaneous apoptosis was achieved by culturing PMNs in RPMI 1640 supplemented with 10% autologous serum, 2 mmol/L glutamine, 100 U/mL penicillin, and 100 µg/mL streptomycin for 20 h at 37 °C in a 5% CO$_2$ atmosphere. Cells were on average 25–50% apoptotic with about 3% necrosis as assessed by light microscopy on stained cytocentrifuged preparations. Differentiated THP-1 cells (5×10^5 cells/well) were exposed to the appropriate stimuli as indicated for 15 min at 37 °C, before co-incubation with apoptotic PMNs (1×10^6 PMNs/well) at 37 °C for 2 h. Non-ingested cells were removed by three washes with cold phosphate-buffered saline. Phagocytosis was assayed by myeloperoxidase staining of co-cultures fixed with 2.5% glutaraldehyde. For each experiment, the number of THP-1 cells containing one or more PMN in at least five fields (minimum of 400 cells) was expressed as a percentage of the total number of THP-1 cells and an average between duplicate wells was calculated.

4.7.14 Cytokine Production by J774 Macrophages

The murine J774 macrophages (European Collection of Cell Cultures, UK) were maintained in suspension of RPMI 1640 supplemented with 2 mmol/L glutamine, 100 IU/mL penicillin, 100 µg/ml streptomycin, and 10% fetal calf serum (Life Technologies Inc, Grand Island, NY). Cells were seeded at 1×10^6 cells per ml for experiments and exposed to lipopolysaccharide (LPS) at a concentration of 100 ng/mL for 24 h at 37 °C in 5% CO$_2$. (1R)-4 and (1S)-4 were added to the cells 1 h before addition of LPS, at a concentration of 1 nM, 1 µM and 10 µM. After 24 h the supernatants were collected for cytokine analysis. IL-1β, MCP and IL-12p40 concentrations in cell culture supernatants were quantified by commercial DuoSet ELISA kits (R&D Systems), according to the manufacturer's instructions.

References

1. O'Sullivan TP, Vallin KSA, Shah STA, Fakhry J, Maderna P, Scannell M, Sampaio ALF, Perretti M, Godson C, Guiry PJ (2007) J Med Chem 50:5894
2. Petasis NA, Keledjian R, Sun Y-P, Nagulapalli KC, Tjonahen E, Yang R, Serhan CN (2008) Bioorg Med Chem Lett 18:1382
3. Sun Y-P, Tjonahen E, Keledjian R, Zhu M, Yang R, Recchiuti A, Pillai PS, Petasis NA, Serhan CN (2009) Prostaglandins Leukotrienes Essent Fat Acids 81:357
4. Maddox JF, Hachicha M, Takano T, Petasis NA, Fokin VV, Serhan CN (1997) J Biol Chem 272:6972
5. Williams DA, Foye WO, Lemke TL (2002) In: Foye's principles of medicinal chemistry, Lippincott Williams & Wilkins, 5th edn, 60
6. Patani GA, LaVoie EJ (1996) Chem Rev 96:3147
7. Parsons ME, Ganellin CR (2006) Br J Pharmac 147:S127
8. http://drugtopics.modernmedicine.com
9. Lind T, Rydberg L, Kylebäck A, Jonsson A, Andersson T, Hasselgren G, Holmberg J, Röhss K (2000) Aliment Pharmacol Ther 14:861
10. Herzig SJ, Howell MD, Ngo LH, Marcantonio ER (2009) J Am Med Assoc 301:2120
11. Colca JR, McDonald WG, Waldon DJ, Leone JW, Lull JM, Bannow CA, Lund ET, Mathews WR (2004) Am J Physiol Endocrinol Metab 286:E252
12. Morii M, Takata H, Fujisaki H, Takegucht N (1990) Biochemical Pharmacol 39:661
13. Duffy CD, Maderna P, McCarthy C, Loscher CE, Godson C, Guiry PJ (2010) Chem Med Chem 5:517
14. Gribble GW, Saulnier MG (1993) Heterocycles 35:151
15. Gilman H, Spatz SM (1951) J Org Chem 16:1485
16. Corey EJ, Suggs JW (1975) Tetrahedron Lett 16:2647
17. Agarwal S, Tiwari HP, Sharma JP (1990) Tetrahedron 46:4417
18. Aoyagi Y, Inariyama T, Arai Y, Tsuchida S, Matuda Y, Kobayashi H, Ohta A, Kurihara T, Fujihira S (1994) Tetrahedron 50:13575
19. Heck RF (1968) J Am Chem Soc 90:5518
20. Nicolaou KC, Bulger PG, Sarlah D (2005) Angew Chem Int Ed Engl 44:4442
21. Coyne AG, Fitzpatrick MO, Guiry PJ (2009) In: Oestreich M (ed) The Mizoroki-Heck Reaction. p 405
22. Labelle M, Belley M, Gareau Y, Gauthier JY, Guay D, Gordon R, Grossman SG, Jones TR, Leblanc Y, McAuliffe M, McFarlane C, Masson P, Metters KM, Ouimet N, Patrick DH, Piechuta H, Rochette C, Sawyer N, Xiang YB, Pickett CB, Ford-Hutchinson AW, Zamboni RJ, Young RN (1995) Bioorg Med Chem Lett 5:283
23. Singh S, Duffy CD, Shah STA, Guiry PJ (2008) J Org Chem 73:6429
24. Zhang Y, Pavlova OA, Chefer SI, Hall AW, Kurian V, Brown LL, Kimes AS, Mukhin AG, Horti AG (2004) J Med Chem 47:2453
25. Robert N, Hoarau C, Celanire S, Ribereau P, Godard A, Queguiner G, Marsais F (2005) Tetrahedron 61:4569
26. Srebnik M, Ramachandran P, Brown H (1988) J Org Chem 53:2916
27. Wu S-H, Liao PY, Dong L, Chen ZQ (2008) Inflamm Res 57:430
28. Decker Y, McBean G, Godson C (2009) Am J Physiol Cell Physiol 296:C1420
29. Aliberti J, Serhan C, Sher A (2002) J Exp Med 196:1253
30. Sodin-Semrl S, Taddeo B, Tseng D, Varga J, Fiore S (2000) J Immunol 164:2660

Chapter 5
Thiophene-Containing Lipoxin A$_4$ Analogues: Synthesis and Their Effect on the Production of Key Cytokines

5.1 Introduction

It has recently been demonstrated that replacement of the triene system, present in native LXA$_4$ and LXB$_4$ with benzene, increases the stability of these eicosanoids to enzymatic metabolism [1–3]. In Chap. 4, we demonstrated that the addition of a heteroatom can also enhance the bioactivity. This pyridine-containing analogue displayed an impressive ability to resolve the inflammation process [4]. In an extension to this work, we sought to replace the triene system with a thiophene ring, Fig. 5.1, and examine the effect this substitution has on the biological potency of the compound. This substitution is a classical example of the bioisoterism concept in medicinal chemistry [5, 6]. Thiophene is an excellent bioisostere for benzene, as the diameter of the sulphur atom is the same length of the replaced C=C double bond [7]. It also offers the possibility of accessing three positional isomers of the analogue which can assist in probing the compounds bioactivity. This particular drug design approach is common and accounts for a large proportion of the successful examples of bioisosteric replacement which can be seen in the literature [8–10].

This bioisosteric replacement has had widespread success in the pharmaceutical industry as it has shown to prevent unwanted side effects in some drugs, which is an ongoing goal throughout the industry. Clozapine, Fig. 5.2, is a powerful drug used to treat schizophrenia and bipolar mania. In spite of the potency of this compound, it is rarely used as a treatment for this condition due to its adverse side effects. These side effects include agranulocytosis [11], autonomic dysregulation and cardiac repolarisation [12]. Replacement of the phenyl group with a substituted thiophene ring dramatically reduced the occurrence of potentially fatal agranulocytosis [11], a condition associated with a dangerously low white blood cell count [13]. Sales of Olanzapine amounted to $1.75 billion in 2008 in the United States alone [14].

C. Duffy, *Heteroaromatic Lipoxin A$_4$ Analogues*, Springer Theses, DOI: 10.1007/978-3-642-24632-6_5, © Springer-Verlag Berlin Heidelberg 2012

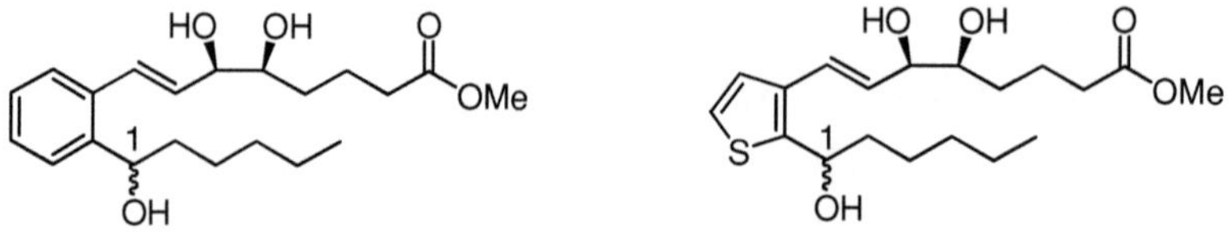

Aromatic LXA$_4$

(1*R/S*)-**1**

Thiophene LXA$_4$

(1*R/S*)-**2**

Fig. 5.1 Design of thiophene-containing LXA$_4$ analogue 2

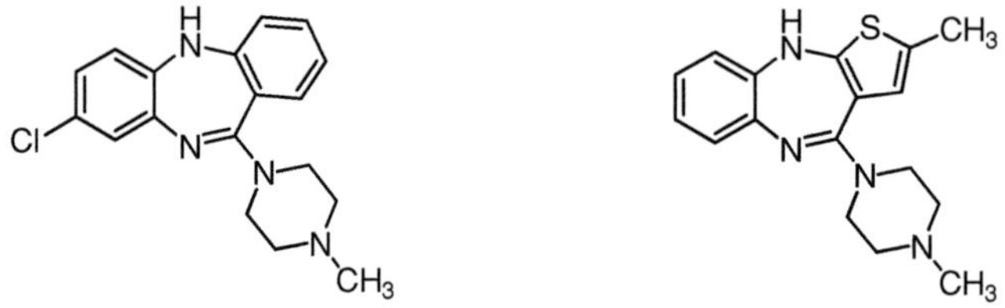

Clozapine

Olanzapine

Fig. 5.2 A successful phenyl/thiophene replacement in medicinal chemistry

5.2 Retrosynthetic Analysis of the Thiophene-Containing LXA$_4$ Analogue

The retrosynthetic analysis of the thiophene-containing LXA$_4$ analogue (1*S*)-**2**, Scheme 5.1, includes an asymmetric reduction of a ketone, a palladium-catalyzed Heck reaction, a Sharpless asymmetric epoxidation, Chap. 2, and a regiospecfic thiophene lithiation.

5.3 Results and Discussion

The initial step in the synthesis requires the preparation of key intermediate **3**, Fig. 5.3, to be employed in a palladium-catalysed Heck reaction.

This compound was prepared in the present study using the readily available starting materials, 3-bromothiophene **5** and hexanal **6**, Scheme 5.2. The procedure employed, reported by Fuller and co-workers, uses an efficient protocol for the deprotonation of bromothiophenes [15]. Using these optimised reaction conditions, 3-bromothiophene **5** was treated with freshly prepared lithium diisopropylamide, formed by the slow addition of *n*-butyllithium to diisopropyl amine in THF at −78 °C. This afforded the corresponding α-lithiated bromothiophene. Quenching this intermediate with hexanal at 0 °C provided alcohol **7** in 75% yield.

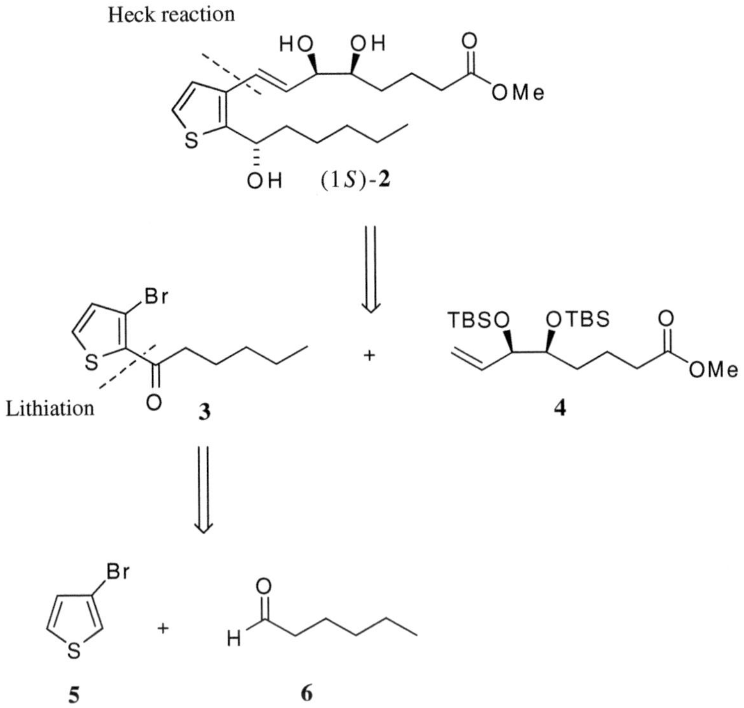
Heck reaction
HO
OH
O
OMe
S
OH
(1S)-2
Br
S
O
Lithiation
3
+
TBSO
OTBS
O
OMe
4
Br
S
+
O
H
5
6

Scheme 5.1 Retrosynthetic analysis of the thiophene-containing LXA_4 analogue (1*S*)-**2**

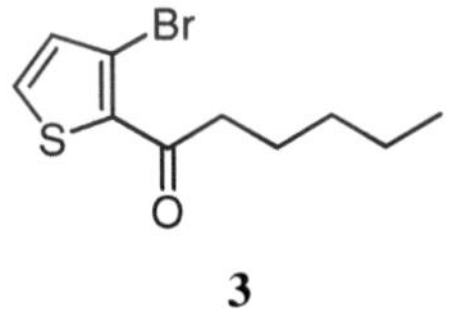
Br
S
O
3

Fig. 5.3 Heck coupling partner ketone **3**

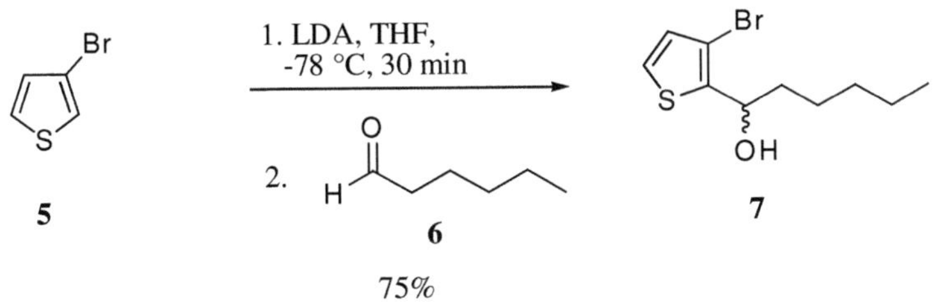
Br
S
5
1. LDA, THF,
-78 °C, 30 min
2.
O
H
6
75%
Br
S
OH
7

Scheme 5.2 Synthesis of alcohol **7**

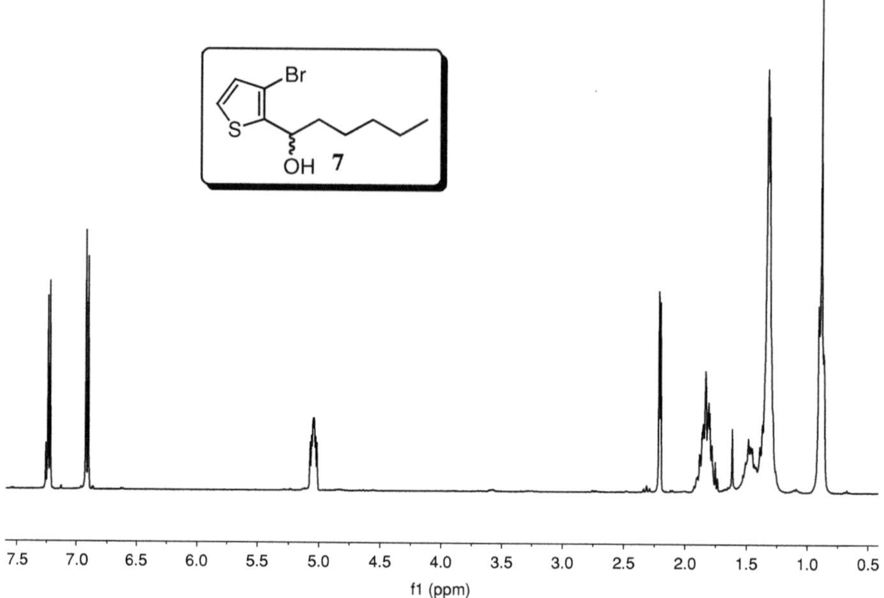

Fig. 5.4 300 MHz ^{1}H NMR spectrum of alcohol **7**

The ^{1}H NMR spectrum of alcohol **7** showed the presence of two doublets in the aromatic region at 7.23 and 6.92 ppm, both integrating for one proton, Fig. 5.4. A broad doublet at 2.22 ppm was also apparent and this signal disappeared with the addition of one drop of D$_2$O indicating the presence of the hydroxyl proton. A signal in the ^{13}C NMR at 69.5 ppm was observed for the CH directly attached to the hydroxyl group. The IR spectrum revealed a characteristic broad hydroxy stretch at 3,348 cm^{-1}.

Alcohol **7** was oxidised using pyridinium chlorochromate in the presence of acetic acid to give ketone **3** in 86% yield, Scheme 5.3 [16, 17]. The ^{1}H NMR of ketone **3** confirmed the formation of the product as a triplet was observed at 3.02 ppm integrating for two protons corresponds to the CH$_2$ directly beside the newly formed carbonyl. A signal at 192.7 ppm in the ^{13}C NMR, along with a sharp stretch at 1,659 cm^{-1} in the IR spectrum, further confirmed the presence of the carbonyl.

Scheme 5.3 Formation of ketone 3

The next step was the construction of the top functionalised alkyl chain. We have previously reported the successful application of the palladium-catalysed Heck reaction for the synthesis of Lipoxin analogues [1, 4]. In light of this, we chose to attempt a Heck coupling between ketone **3** and olefin **4**. Employing the same reaction conditions which were used for our benzene-containing LXA$_4$ analogues, the *trans* olefin was formed in 75% yield, Scheme 5.4 [1]. The reaction involved the use of palladium acetate (10 mol%) and tri-*o*-tolyphosphine with tributylamine as the solvent and the base. After 24 h, Heck coupled intermediate **8** was isolated as the sole product.

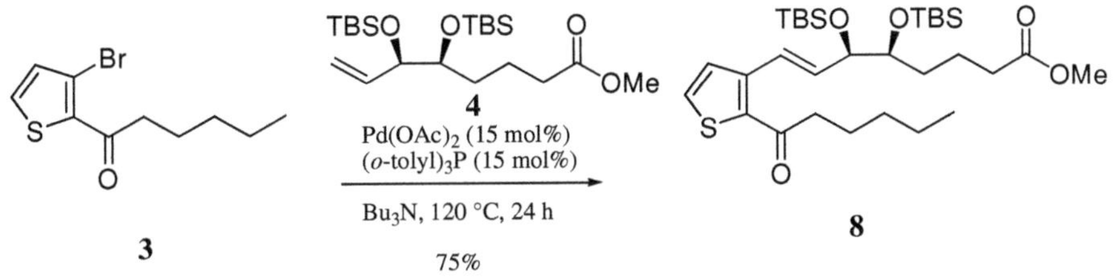

Scheme 5.4 Palladium-catalysed Heck reaction

The ^{1}H NMR spectrum of ketone **8** shows a doublet at 7.43 ppm and a double doublet at 6.19 ppm with a large coupling constant of 16.2 Hz confirming the required *E*-stereochemisty had been achieved, Fig. 5.5. The ^{13}C NMR spectrum of ketone **8** also contains distinct olefin carbon signals at 130.2 and 140.5 ppm. Two sharp stretches also appeared in the IR spectrum at 1,741 and 1,668 cm^{-1} for the ester and ketone carbonyls, respectively.

This ketone was subsequently reduced using Brown's (−)-β-chlorodiisopinocampheylborane to afford (*S*)-alcohol **9** in 49% yield, Scheme 5.5 [18]. A multiplet at 5.09 ppm in the ^{1}H NMR spectrum integrating for one proton indicated the reduction was successful. Only one sharp stretch at 1,741 cm^{-1} remained in the IR spectrum, corresponding to the ester carbonyl vibrational stretch. A *de* of 94% for alcohol **9** was determined by chiral HPLC.

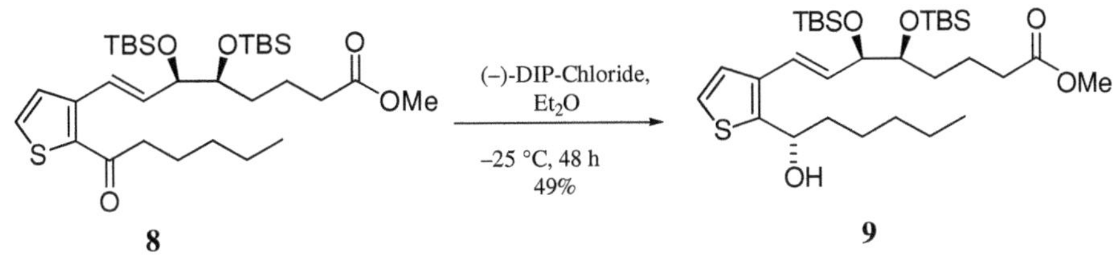

Scheme 5.5 Asymmetric reduction of ketone **8** affording alcohol **9**

Removal of the silyl ether protecting groups proved to be extremely difficult in this case. Conditions used for the deprotection of the previously reported benzene- and pyridine-containing LXA$_4$ analogues [1, 4], failed to furnish the required

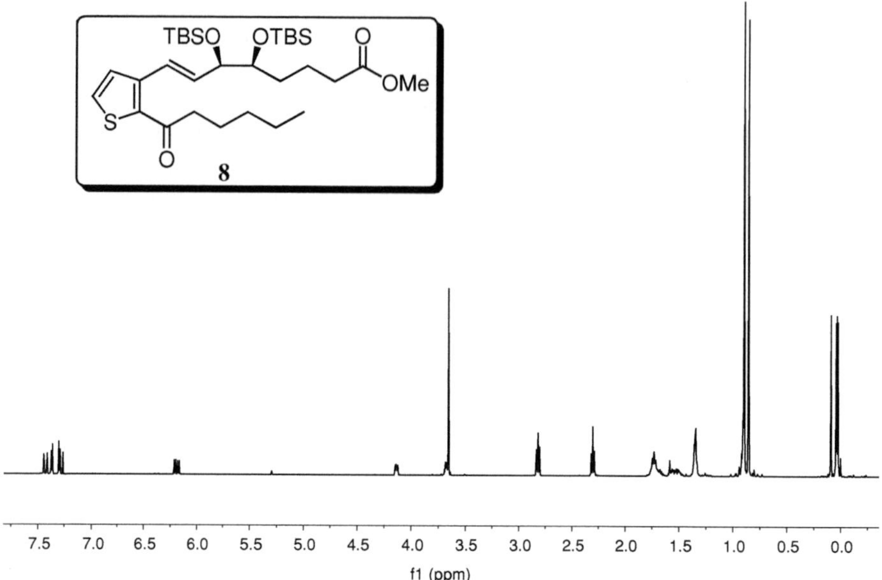

Fig. 5.5   500 MHz ^{1}H NMR spectrum of ketone **8**

product as did the use of TMSBr or ZrCl$_4$ in MeOH. Exposure of **9** to a variety of known deprotection conditions including I$_2$/MeOH, HCl/EtOH and TBAF/THF/ 4Å molecular sieves predominantly led to decomposition of the product [19, 20]. A range of other deprotection protocols including HCOOH/THF/H$_2$O and PPTS/ EtOH, had no effect on compound **9**, with only starting material being recovered [21, 22]. These failed deprotection reactions are probably due to dehydration to form a benzylic carbocation which is stabilised by resonance.

An attempted deprotection of ketone **8** using TBAF in THF proved successful. However, under these conditions, a 1:1 mixture of product **10** and by-product lactone **11** was formed, Scheme 5.6.

Scheme 5.6   Deprotection of ketone **8**

Compounds **10** and **11** have a similar retention factor on silica gel chromatography making them very difficult to separate and hence making purification a challenge. A solvent system of 95:5 CH_2Cl_2:methanol resulted in minimal separation and allowed **10** to be isolated using preparative TLC. After purification ketone **10** was then further protected using 2,2-dimethoxypropane and reduced with NaBH$_4$—in methanol providing (1*R/S*)-**13** in 68% yield, Scheme 5.7.

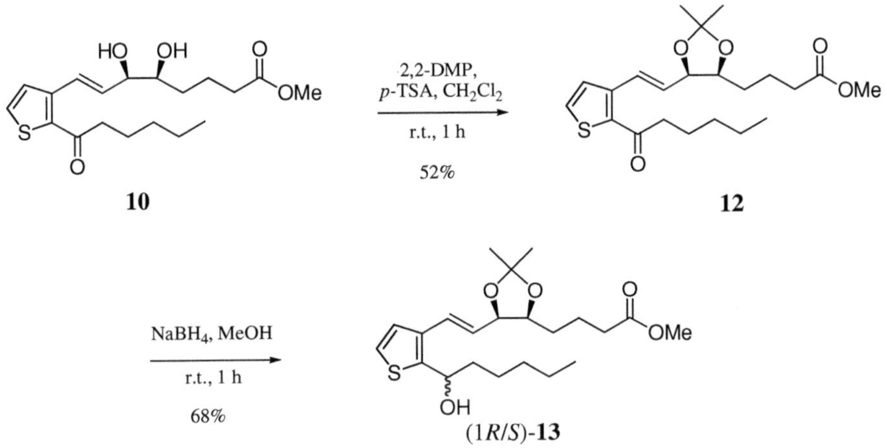

Scheme 5.7 Synthesis of (1*R/S*)-**13** from diol **10**

Evidence for the formation of epimeric **13** could be seen by analysis of its ^{1}H NMR spectrum which contained a multiplet at 5.08 ppm corresponding to the CH adjacent to the hydroxyl group, Fig. 5.6. Two distinct methyl ester peaks were also present at 3.63 and 3.61 ppm.

Efforts then focused on the deprotection of alcohol (1*R/S*)-**13** using 2 N HCl in THF, Scheme 5.8. The reaction was monitored by TLC over a period of 1.5 h and the appearance of the product was apparent. The reaction was stopped, purified and analysed by ^{1}H NMR spectroscopy. The ^{1}H NMR spectrum confirmed the formation of product **2**. However, upon removal of the solvent at low temperature, the product decomposed, turning black in colour. This decomposition was also evident in the ^{1}H NMR spectrum.

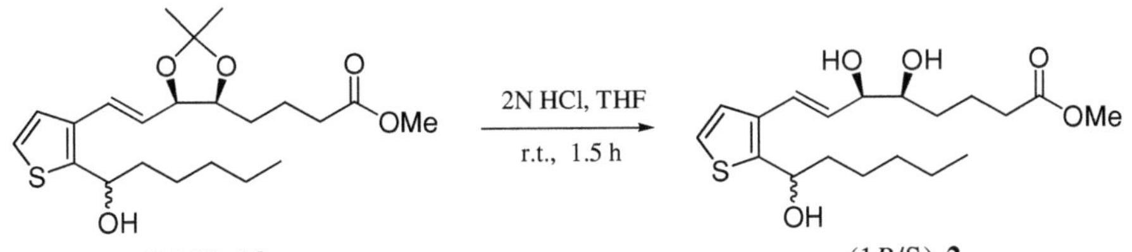

Scheme 5.8 Attempted deprotection of epimeric alcohol (1*R/S*)-**13**

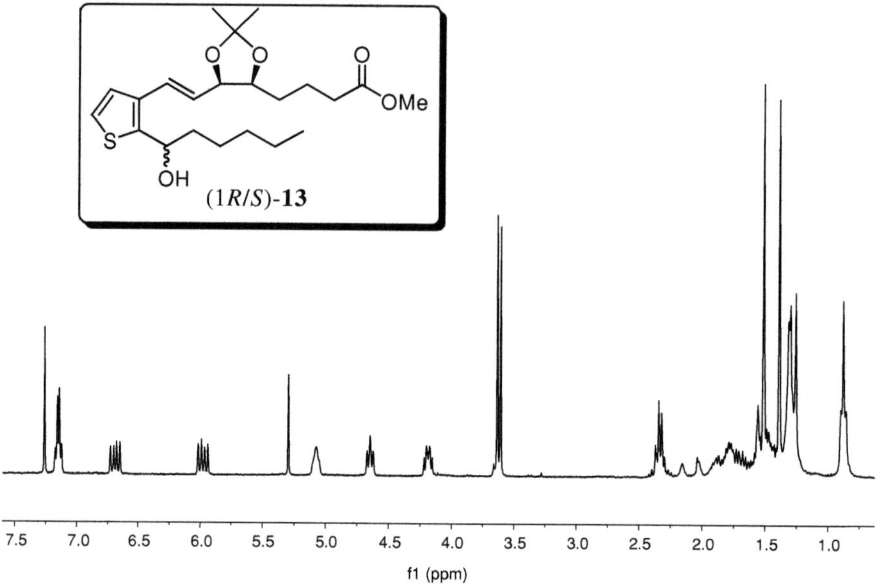

Fig. 5.6 300 MHz ^{1}H NMR spectrum of alcohol (1*R/S*)-**13**

5.4 Protecting Group-Free Synthesis of the Thiophene-Containing LXA$_4$

At this point, a protecting group-free synthesis of the thiophene-containing LXA$_4$ **2** was attempted. This new synthesis relies on the formation of the *trans* olefin via Grubbs' cross metathesis, Scheme 5.9. Grubbs' cross metathesis reaction

Scheme 5.9 Retrosynthetic analysis of protecting group-free synthetic route

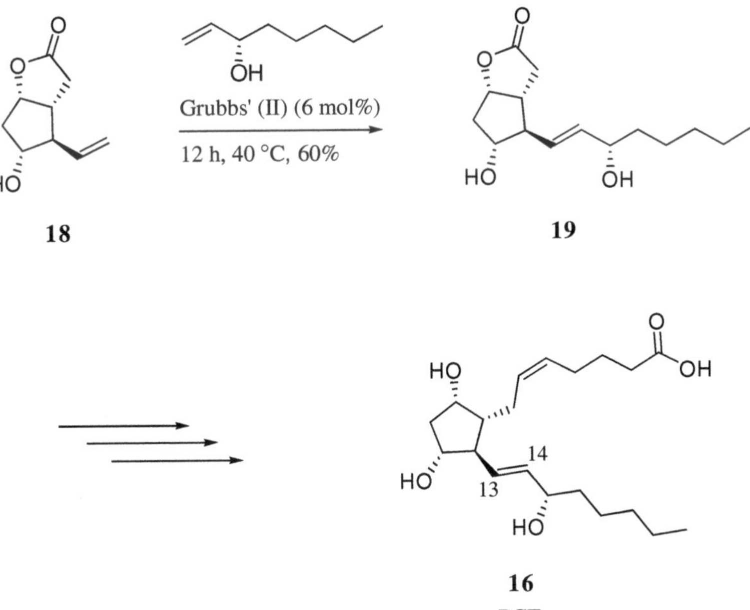

Fig. 5.7 Prostaglandins F$_{2\alpha}$ **16** and J$_2$ **17**

conditions usually tolerate a wide range of functional groups [23–25]. This alternative route would have the added advantage of reducing the overall number of synthetic steps, therefore providing a more economical synthetic route.

A similar approach was taken by Sheddan and Mulzer in their synthesis of prostaglandins F$_{2\alpha}$ **16** and J$_2$ **17**, Fig. 5.7 [26, 27]. The *trans* olefin at C$_{13-14}$ on the ω-side chain was constructed via Grubbs' cross metathesis.

The *trans* olefin in prostaglandins F$_{2\alpha}$ **16** was synthesised by the coupling of bicylic olefin **18** with an allylic alcohol, Scheme 5.10. Grubbs' second generation catalyst was employed providing **19** in 60% yield. Improved yields of 84% were accomplished when all hydroxyl groups were protected as their silyl ethers.

Scheme 5.10 Synthesis of prostaglandin F$_{2\alpha}$ **16**

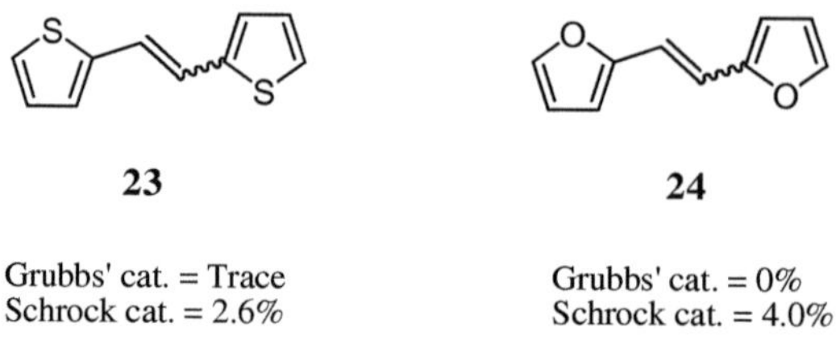

23

24

Grubbs' cat. = Trace   Grubbs' cat. = 0%
Schrock cat. = 2.6%   Schrock cat. = 4.0%

Fig. 5.8 Potential homodimerisation formed from 2-vinyl heterocycles

Kawai et al. [28] have also demonstrated successful Grubbs' cross metathesis reactions between 2-vinythiophene **21** and 1-octene **22**, Scheme 5.11.

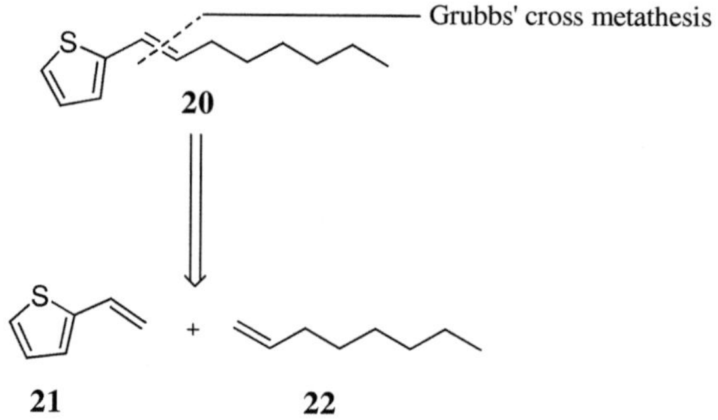

Grubbs' cross metathesis

20

21   **22**

Scheme 5.11 Cross metathesis of 2-vinythiophene **21** and 1-octene **23** [28]

The authors also showed the reaction proceeds with 2-vinylfuran and 1-octene **22**. They also provide an in-depth study of the activity of both Grubbs' and Schrock catalysts in the self-metathesis reactions of the vinyl heterocycles. It was observed that very little homodimerisation, to form **23** and **24**, occurred with both catalysts, Fig. 5.8.

The synthesis of alcohol **14** in the present study relies on a palladium-catalysed Stille coupling reaction of bromide **3**, Scheme 5.12. This is a powerful method for the cross coupling of aryl bromides and organotin compounds [29]. Reactions conditions employed for a related synthesis [30, 31], rely on the use of Pd(PPh$_3$)$_4$ (10 mol%), tributyl(vinyl)stannane, LiCl and 1,4-dioxane as the solvent at 80 °C

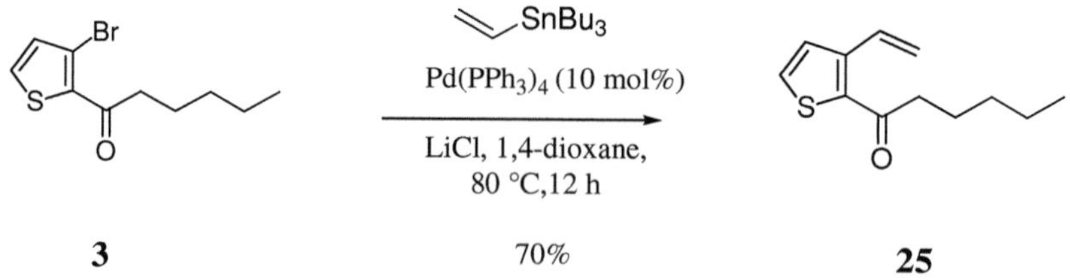

SnBu$_3$

Pd(PPh$_3$)$_4$ (10 mol%)

LiCl, 1,4-dioxane,
80 °C,12 h

3   70%   **25**

Scheme 5.12 Stille reaction conditions used to synthesise **25** [30, 36]

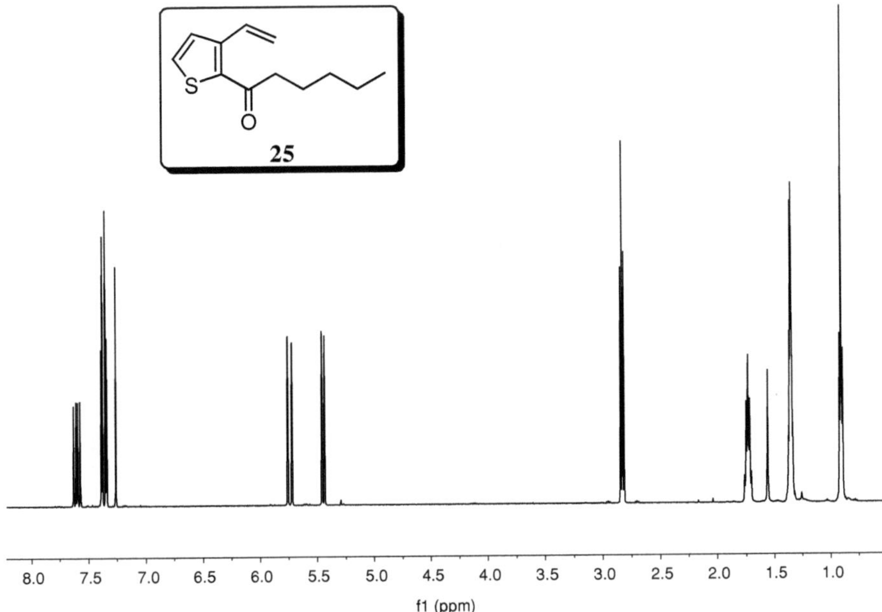

Fig. 5.9   500 MHz ^{1}H NMR spectrum of ketone 25

for 12 h. Exploiting these reaction conditions gave the required Stille coupled product ketone **25** in 70% isolated yield.

A slightly lower yield of 65% was obtained when the reaction was performed using microwave irradiation. This reduced the reaction time from 12 h at 80 °C to 45 min at 120 °C.

The formation of the ketone **25** was evident by the appearance of three double doublets in the ^{1}H NMR spectrum at 7.57, 5.71 and 5.41 ppm integrating for one proton each, Fig. 5.9. A triplet at 2.80 ppm integrating for two protons revealed the presence of the CH$_2$ next to the carbonyl. The vinyl carbons signals were observed at 131.1 and 118.4 ppm in the ^{13}C NMR spectrum. A characteristic carbonyl stretch was observed at 1,664 cm^{-1} in the IR spectrum confirming the presence of the ketone moiety.

Ketone **25** was then reduced using sodium borohydride in MeOH at room temperature to give the racemic alcohol **14** in 63% yield, Scheme 5.13.

Scheme **5.13**   Reduction of ketone 25 using sodium borohydride in MeOH

5.5 Attempted Grubbs' Cross Coupling Reaction

With this in hand, we attempted the cross coupling of vinyl alcohol **14** with some commercially available terminal alkenes to analyse if any cross coupling would occur. Three catalysts; Grubbs' 1st gen., Grubbs 2nd gen., and Hoveyda–Grubbs cat. were screened, Fig. 5.10.

The optimum reactions conditions featured Grubbs 2nd gen. catalyst in dichloromethane at 40 °C. The cross coupling of alcohol **14** with the commercially available olefins resulted in the successful isolation of products which were solely analysed by ^{1}H NMR spectroscopy on small scale, Fig. 5.11.

This inspired the synthesis of an asymmetric alcohol (1*R*)-**14**, in the hope that this intermediate could also be applied in a successful cross metathesis reaction. Ketone **25** was reduced using Brown's (+) chlorodiisopinocampheylborane [18], but was extremely difficult to purify form a side product of the reaction. Therefore ketone **25** was reduced using the CBS method, developed by Corey [31], which gave alcohol (1*R*)-**14** in 65% yield, Scheme 5.14.

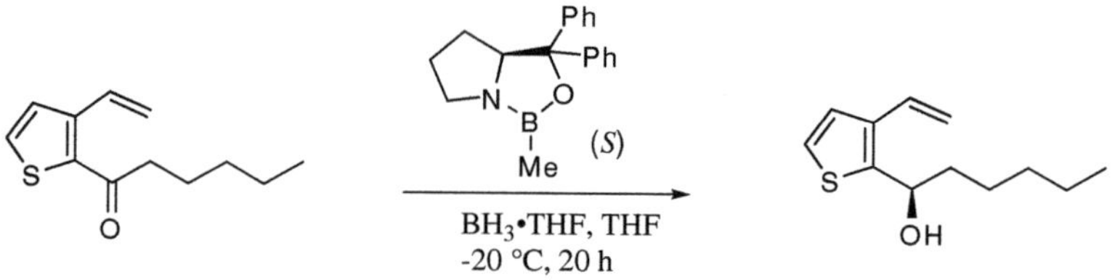

Scheme 5.14 CBS-catalysed reduction of ketone **25**

The ^{1}H NMR spectrum of the newly formed alcohol (1*R*)-**14** contained a multiplet at 5.09 ppm integrating for one proton. ^{1}H NMR spectroscopic evidence, coupled with the disappearance of the carbonyl stretch in the IR spectrum, indicated the reduction had taken place. The hydroxy proton resonated as a broad doublet in the ^{1}H NMR spectrum, as a consequence of it coupling to the newly formed CH. This doublet disappeared with the addition of D$_2$O. The newly formed CH was visible at 68.4 ppm in the ^{13}C NMR spectrum. A broad stretch at 3,358 cm^{-1} was also observed in the IR spectrum. An *ee* value of 94% for alcohol (1*R*)-**14** was determined by chiral HPLC, Fig. 5.12. This *ee* value drops to 85% when the reaction was performed at room temperature.

The cross metathesis of olefin **15** and alcohol (1*R*)-**14** was attempted using Grubbs 1st and 2nd generation catalysts as well as the Hoveyda–Grubbs' catalyst, Scheme 5.15. However, under these conditions only starting material was recovered and no coupled product was observed.

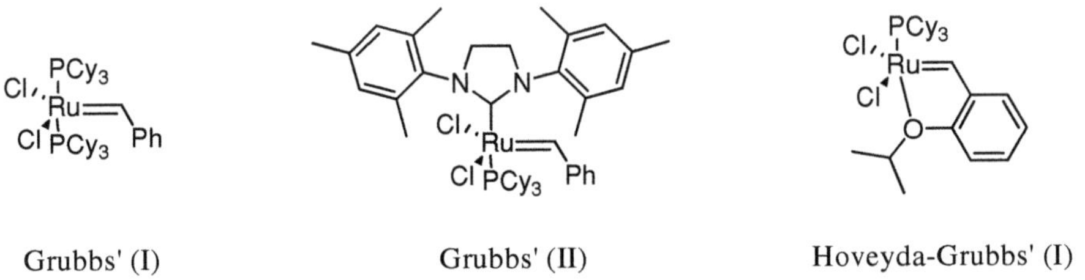

Grubbs' (I) Grubbs' (II) Hoveyda-Grubbs' (I)

Fig. 5.10 Three catalysts screened; Grubbs' 1st gen., Grubbs 2nd gen., and Hoveyda–Grubbs cat

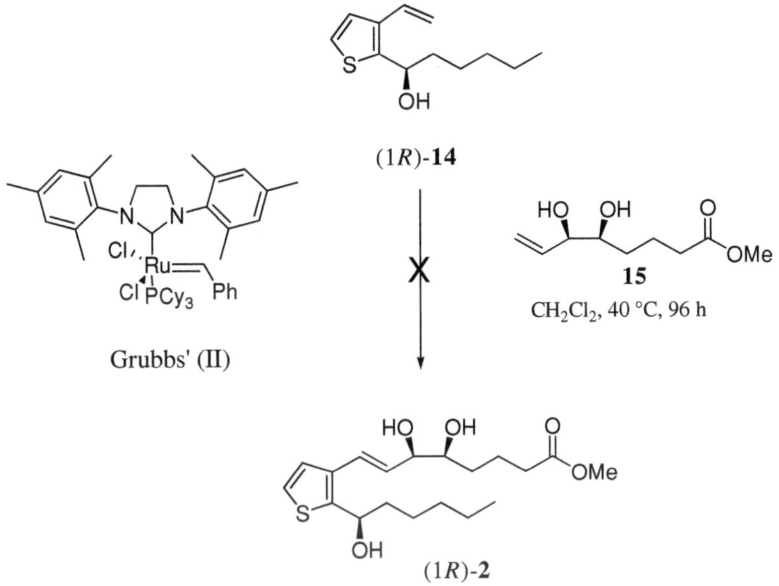

$(1R)$-**14**

Grubbs' (II)

15

CH_2Cl_2, 40 °C, 96 h

$(1R)$-**2**

Scheme 5.15 Attempted cross metathesis of alcohol $(1R)$-**14** and olefin **15**

At this stage a decision was made to protect the olefin **15** with 2, 2-dimethoxypropane. This protected olefin **26** was prepared from the corresponding diol **15**, whose synthesis was discussed in detail in Chap. 3. Use of 2,2-dimethoxypropane and *p*-TSA in dichloromethane afforded **26** in 75% yield, Scheme 5.16.

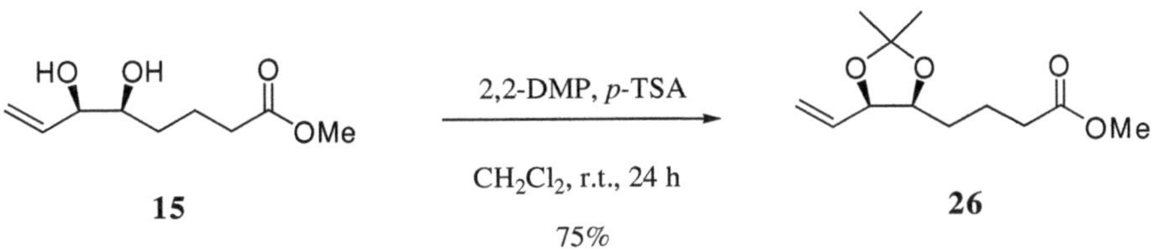

15

2,2-DMP, *p*-TSA

CH_2Cl_2, r.t., 24 h

75%

26

Scheme 5.16 Protection of diol **15** using 2,2-dimethoxypropane and *p*-TSA

Olefin **26** and alcohol (1*R*)-**14** were reacted together in a cross metathesis reaction using the conditions described above and produced compound (1*R*)-**13** in 32% yield, Scheme 5.17.

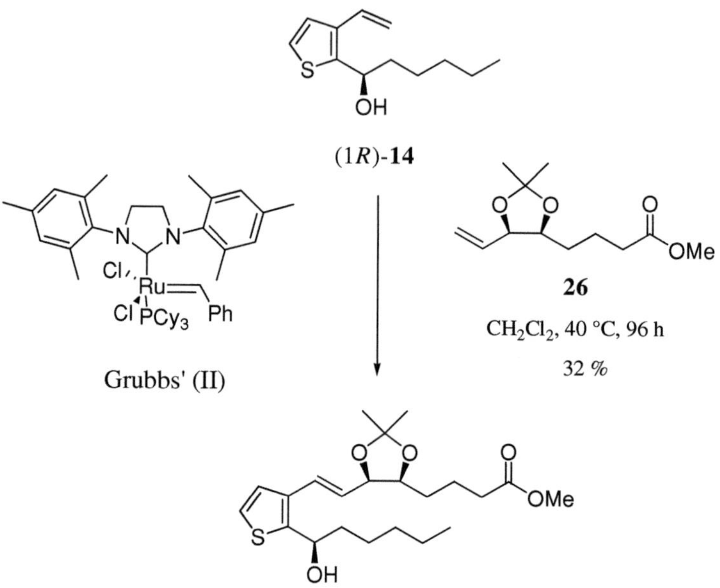

(1*R*)-**14**

26

Grubbs' (II)

CH$_2$Cl$_2$, 40 °C, 96 h

32 %

(1*R*)-**13**

Scheme 5.17 Cross metathesis providing (1*R*)-**13**

Following the successful synthesis of (1*R*)-**13**, a deprotection was now attempted using 2N HCl. This again proved to be unsuccessful as the product decomposed during the work up.

5.6 Biological Evaluation

In a continuation of our efforts to find enzymatically stable Lipoxin analogues, we turned our attention towards screening intermediates **10** and (1*R*)-**13**, Fig. 5.13.

These intermediates were screened for their ability effect the production of key cytokines, a characteristic of the native LX [32–35] and also the pyridine-containing LXA$_4$, Chap. 4.

Compound **10** promoted the production of IL-12p40 at low concentrations, Fig. 5.14, while compound (1*R*)-**13** decreased the production but only at high concentrations.

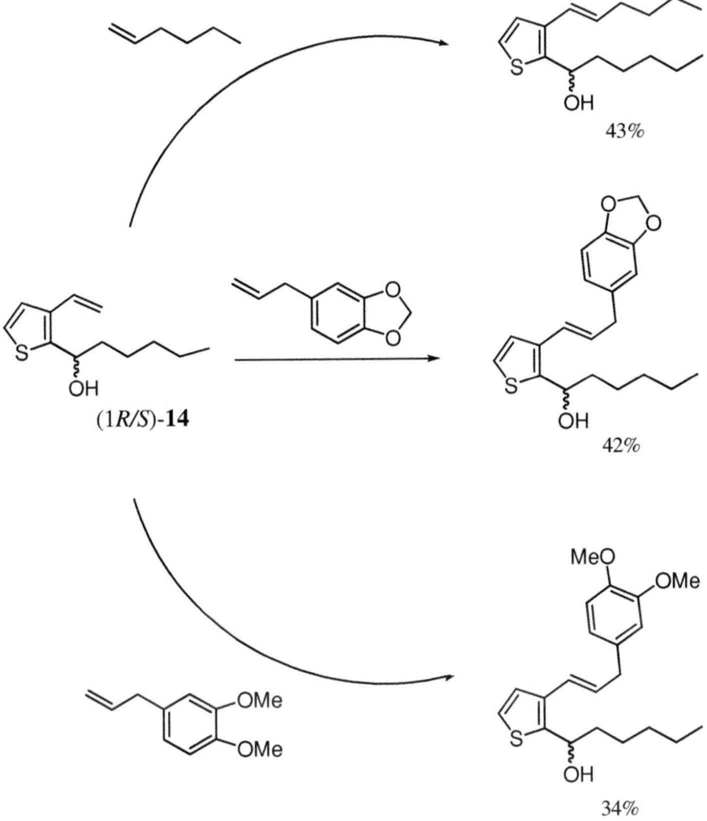

Fig. 5.11 Trial cross metathesis reactions carried out with Grubbs' 2nd gen. cat., CH$_2$Cl$_2$, at 40 °C for 5 days

Compound **10** effected the production of TNF-alpha with an increase at low concentrations, Fig. 5.15. Compound (1R)-**13** had no effect.

No effect was observed with compound **10** on the production IL-1 beta, Fig. 5.16, although compound (1R)-**13** caused an increase of this cytokine at low concentrations.

Compound **10** and (1R)-**13** decreased the production of IL-6 at relatively high concentrations, Fig. 5.17.

Both compounds **10** and (1R)-**13** caused a small decrease in the production of Monocyte Chemoattractant Protein (MCP) at low concentrations, Fig. 5.18.

A decrease was observed in the production of Macrophage Inflammatory Protein-1 (MIP-1) alpha and Macrophage Inflammatory Protein-2 (MIP-2), Figs. 5.19 and 5.20, respectively.

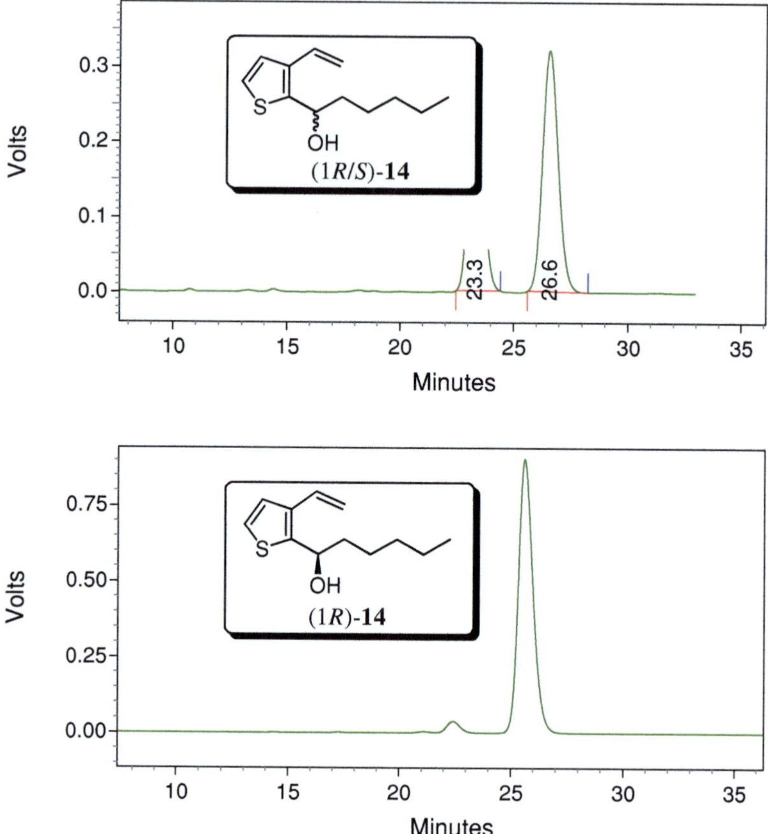

Fig. 5.12 HPLC traces of racemic and enantiopure alcohol 14 performed on a Chiracel® OD column 99:1 hexane/2-propanol, 1.0 mL/min, t$_R$ = 23.3 min for (S), t$_R$ = 26.6 min for (R)

10   (1R)-**13**

Fig. 5.13 Intermediates 12 and (1R)-15

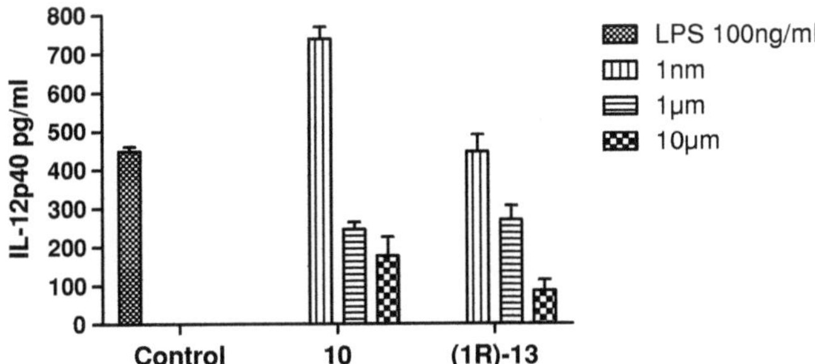

Fig. 5.14 Effect of **10** and(1*R*)-**13** on IL-12p40

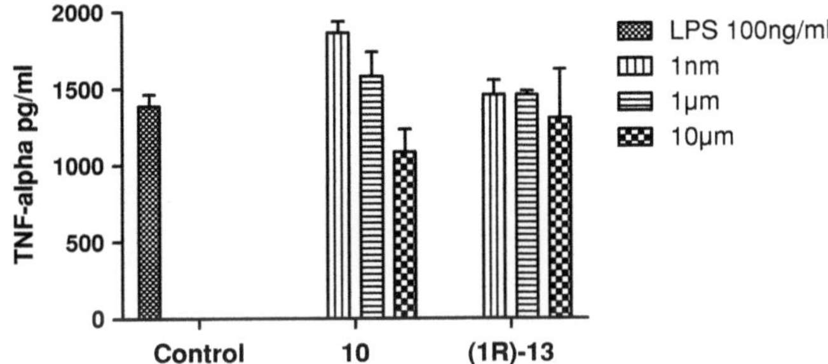

Fig. 5.15 Effect of **10** and (1*R*)-**13** on TNF-alpha

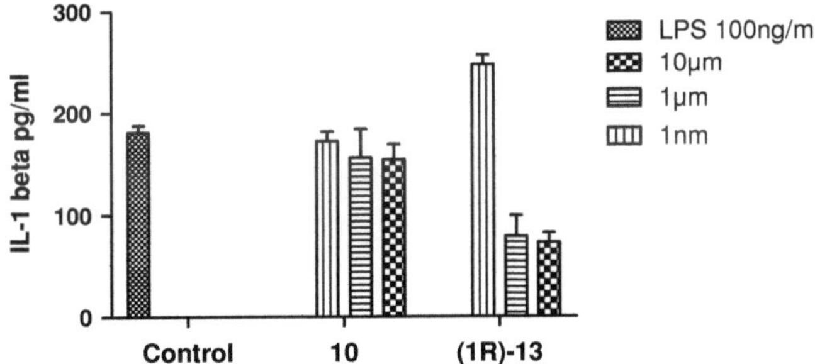

Fig. 5.16 Effect of **10** and (1*R*)-**13** on IL-1 beta

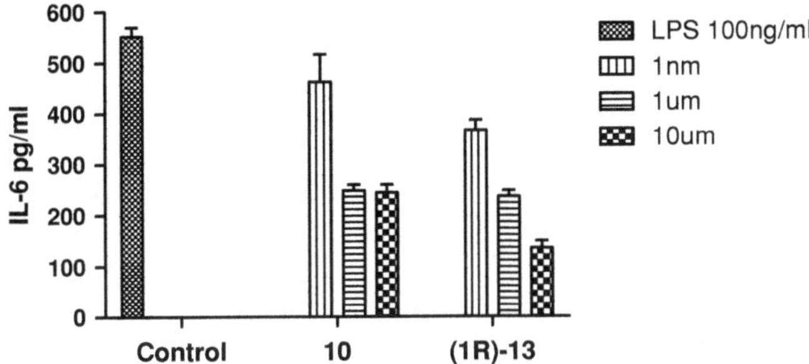

Fig. 5.17 Effect of **10** and (1*R*)-**13** on IL-6

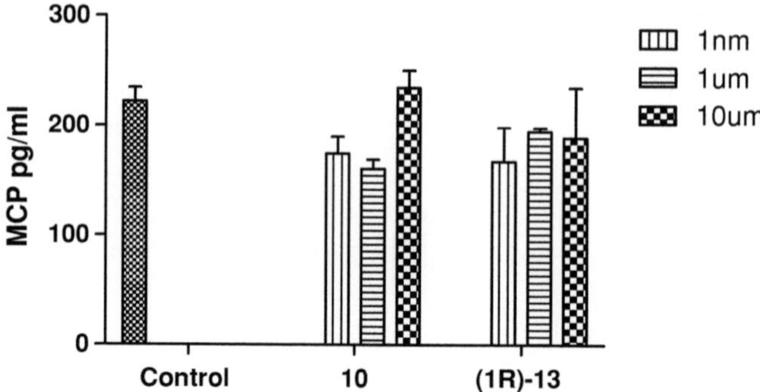

Fig. 5.18   Effect of 10 and (1*R*)-13 on MCP

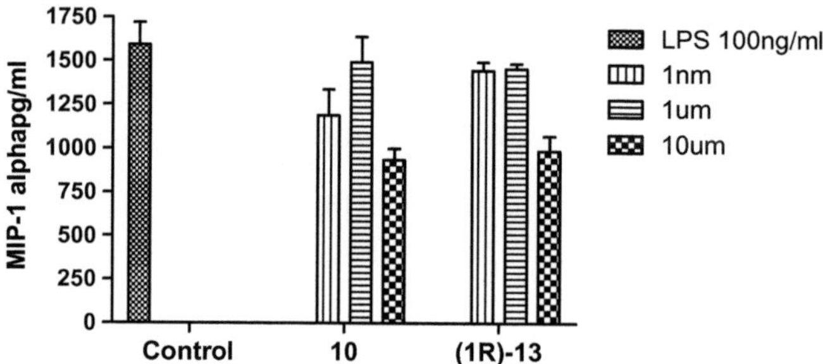

Fig. 5.19   Effect of **10** and (1*R*)-**13** on MIP-1

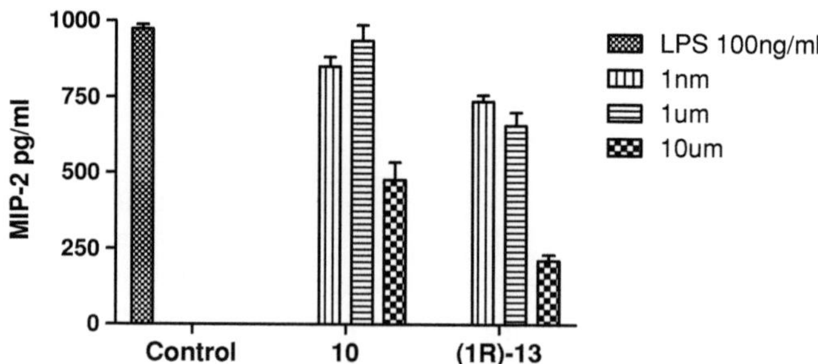

Fig. 5.20   Effect of **10** and (1*R*)-**13** on IL-1 beta

5.7 Conclusion

In summary, our continuing interest in the synthesis of stable LXA_4 analogues, led us to design a reterosynthetic approach for the construction of a thiophene-containing analogue. Our inability to deprotect the diol functionality in the final step prompted us to attempt a protecting group free synthesis. Parallel to our ongoing synthetic efforts, we have shown that these novel intermediates play an important role in the promotion or reduction of important cytokines. These encouraging results have motivated further investigations directed towards the synthesis of this thiophene-containing LXA_4 analogue.

5.8 Experimental

5.8.1 1-(3-Bromothiophen-2-yl)hexan-1-ol (7)

Br
1. LDA, THF
-78 °C, 30 min
2.
O
H
6
75%
Br
S
OH
7
S
5

3-Bromothiophene **5** (1.04 mL, 10.9 mmol) was added dropwise to a stirred solution of lithium diisopropylamide prepared by addition of butyllithium (4.4 mL, 2.5 M in hexane; 10.96 mmol) to diisopropylamine (1.54 mL, 10.9 mmol) in tetrahydrofuran (30 mL) at 0 °C and the resulting mixture was stirred for a further 30 min at this temperature prior to addition of hexanal **6** (1.44 mL, 12.1 mmol). The mixture was stirred for 1 h at 0 °C, quenched with saturated aqueous ammonium chloride (50 mL) and extracted with diethyl ether (3 × 50 mL). The combined extracts were washed with brine (50 mL) and dried over Na_2SO_4. The solvent was evaporated and the resulting oil was purified by silica gel column chromatography (pentane/ethyl acetate, 12:1) to afford **7** (2.15 g, 75%) as a colourless oil; TLC: $R_f = 0.46$ (pentane/ethyl acetate, 9.5:0.5); ^{1}H NMR (300 MHz, $CDCl_3$) δ 7.23 (d, $J = 5.3$ Hz, 1H), 6.91 (d, $J = 5.3$ Hz, 1H), 5.05 (m, 1H), 2.22 (br s, 1H), 1.83 (m, 2H), 1.17–1.57 (m, 6H), 0.89 (m, 3H) ppm; ^{13}C NMR (75 MHz, $CDCl_3$) δ 143.3, 129.8, 124.6, 107.6, 69.5, 38.4, 31.5, 25.3, 22.5, 14.0 ppm; IR (neat) (v_{max}, cm^{-1}) 3348, 2929, 1458, 874; HRMS (EI) Found 262.0025 [M], $C_{10}H_{15}BrOS$ requires 262.0027.

5.8.2 1-(3-Bromothiophen-2-yl)hexan-1-one (3)

Glacial acetic acid (0.5 mL) was added to a vigorously stirred solution of pyridinium chlorochromate (2.23 g, 10.6 mmol) in dry dichloromethane (50 mL). After 5 min at room temperature, alcohol **7** (1.85 g, 7.0 mmol) in dichloromethane (5 mL) was added and the resulting mixture was stirred for 5 h. Diethyl ether (100 mL) was then added and the mixture filtered twice. The resulting mixture was concentrated and purified by silica gel column chromatography (pentane/ethyl acetate, 15:1) to afford **3** (1.58 g, 85%) as a colourless oil. TLC: R$_f$ = 0.69 (pentane/ethyl acetate, 9:1) ^{1}H NMR (300 MHz, CDCl$_3$) δ 7.49 (d, J = 5.3 Hz, 1H), 7.10 (d, J = 5.3 Hz, 1H), 3.02 (t, J = 7.5 Hz, 2H), 1.75 (m, 2H), 1.38 (m, 4H) 0.92 (m, 3H) ppm; ^{13}C NMR (75 MHz, CDCl$_3$) δ 192.7, 138.6, 133.6, 131.6, 113.7, 41.5, 31.4, 23.9, 22.5, 13.9 ppm; IR (neat) (v_{max}, cm^{-1}) 2956, 1659, 1408; HRMS (ESI) Found 260.9955 [M + H]$^+$, C$_{10}$H$_{14}$BrOS requires 260.9949.

5.8.3 (5S,6R,E)-Methyl 5,6-bis(tert-butyldimethylsilyloxy)-8-(2-hexanoylthiophen-3-yl)oct-7-enoate (8)

Pd(OAc)$_2$ (20 mg, 0.089 mmol) and P(*o*-tolyl)$_3$ (30 mg, 0.099 mmol) were dissolved in Bu$_3$N (2.5 mL) and stirred at room temperature for 10 min under nitrogen. Bromide **3** (235 mg, 0.9 mmol) was added followed by olefin **4** (250 mg, 0.6 mmol) and the reaction mixture was stirred in a sealed tube at 120 °C for 24 h. The resulting mixture was filtered through a pad of silica gel, eluted with diethyl ether (100 mL) and the solvent was removed in vacuo. The remaining Bu$_3$N was removed by Kugelrohr distillation at 100 °C. The residue was purified by silica gel column chromatography (pentane:ethyl acetate, 20:1) to afford **8** (268 mg, 75%) as a yellow viscous oil; TLC: R$_f$ = 0.51 (pentane/diethyl ether, 10:1) [α]$_D^{20}$ −17.5

($c = 1$, CHCl$_3$); ^{1}H NMR (300 MHz, CDCl$_3$) δ 7.43 (d, $J = 16.2$ Hz, 1H), 7.37 (d, $J = 5.3$ Hz, 1H), 7.29 (d, $J = 5.3$ Hz, 1H), 6.19 (dd, $J = 16.2$, 7.5 Hz, 1H), 4.13 (m, 1H), 3.68 (m, 1H), 3.66 (s, 3H), 2.82 (t, $J = 7.5$ Hz, 2H), 2.30 (t, $J = 7.2$ Hz, 2H), 1.22–1.84 (m, 10H), 0.91 (m, 3H) 0.90 (s, 9H), 0.86 (s, 9H), 0.00–0.09 (m, 12H) ppm; ^{13}C NMR (75 MHz, CDCl$_3$) δ 198.6, 178.7, 148.4, 140.5, 139.9, 133.9, 132.2, 130.2, 56.1, 47.1, 39.0, 37.7, 36.1, 30.7, 30.6, 29.0, 27.2, 25.3, 22.9, 22.8, 18.6, 0.8, 0.7, 0.1, 0.0 ppm; IR (neat) ($v_{\max}$, cm^{-1}) 2954, 1741, 1668, 1414; HRMS (ESI) Found 619.3272 [M + Na]$^+$ C$_{31}$H$_{56}$O$_5$SSi$_2$Na requires 619.3285.

5.8.4 (5S,6R,E)-Methyl-5,6-bis(tert-butyldimethylsilyloxy)-8-(2-((S)-1-hydroxyhexyl)thiophen-3-yl)oct-7-enoate (9)

To a solution of (–)-ß-chlorodiisopinocampheylborane (85 mg; 0.26 mmol) in dry diethyl ether (1 mL) at –20 °C under nitrogen was added ketone **8** (111 mg; 0.19 mmol) in dry diethyl ether (1.5 mL). The solution was stirred at –20 °C for 36 h, allowed to warm to room temperature, and diluted with diethyl ether (1 mL) and pentane (1 mL). Diethanolamine (55 mg; 0.52 mmol) was then added and the resulting mixture was stirred at room temperature for 3 h. Filtration followed by evaporation of the solvent gave an oil which was purified by silica gel column chromatography (pentane:ethyl acetate, 15:1 then, 10:1) to yield **9** (54 mg; 49%) as a viscous colourless oil. $de = 94\%$ as determined by chiral HPLC using a OD column (hexane: iPrOH, 99:1, flow rate: 0.5 mL/min), 13.9 min for (S) and 15.9 min for (R); $[\alpha]_D^{20}$ –18.1 ($c = 1$, CHCl$_3$); TLC: R$_f$ = 0.29 (pentane/ethyl acetate, 15:1); ^{1}H NMR (300 MHz, CDCl$_3$) δ 7.16 (d, $J = 5.3$ Hz, 1H), 7.10 (d, $J = 5.3$ Hz, 1H), 6.54 (d, $J = 15.9$ Hz, 1H), 6.01 (dd, $J = 15.9$, 6.9 Hz, 1H), 5.09 (m, 1H), 4.10 (dd, $J = 6.3$, 4.5 Hz, 1H), 3.66 (m, 1H), 3.65 (s, 3H), 2.30 (dt, $J = 7.2$, 1.5 Hz, 2H), 1.17–1.98 (m, 12H), 0.91 (m, 3H) 0.90 (s, 9H), 0.87 (s, 9H), 0.00–0.08 (m, 12H) ppm; ^{13}C NMR (75 MHz, CDCl$_3$) δ 178.7, 148.9, 140.0, 135.8, 130.0, 128.3, 127.8, 72.8, 56.1, 44.1, 38.9, 37.7, 36.2, 30.6, 30.2, 27.2, 27.1, 25.1, 22.9, 22.7, 18.6, 0.6, 0.5, 0.1, 0.0 ppm; IR (neat) ($v_{\max}$, cm^{-1}) 3452, 2933, 1741, 1253; HRMS (ESI) Found 621.3468 [M + Na]$^+$, C$_{31}$H$_{58}$O$_5$SSi$_2$Na requires 621.3441.

5.8.5 (5S,6R,E)-Methyl 8-(2-hexanoylthiophen-3-yl)-5,6-dihydroxyoct-7-enoate (10)

Ketone **8** (143 mg, 0.239 mmol) was dissolved in dry THF (1.5 mL) to which a TBAF (1 M in THF, 1.31 mL, 1.31 mmol) was added slowly under an atmosphere of N$_2$. The reaction mixture was stirred for 1.5 h. The solvent was removed in vacuo and the residue was purified by preparative silica gel TLC (CH$_2$Cl$_2$:MeOH, 95:5) to afford **10** as a viscous colourless oil (27 mg, 30% yield). TLC: R$_f$ = 0.42 (CH$_2$Cl$_2$/MeOH, 9.5:0.5); [α]$_D^{20}$ + 1.6 (c = 0.98, CHCl$_3$), ^{1}H NMR (500 MHz, CDCl$_3$) δ ppm 7.55 (d, J = 16.2 Hz, 1H), 7.38 (d, J = 5.1 Hz, 1H), 7.33 (d, J = 5.1 Hz, 1H), 6.32 (dd, J = 16.2, 7.2 Hz, 1H), 4.28 (m, 1H), 3.79 (m, 1H), 3.66 (s, 3H), 2.83 (t, J = 7.3 Hz, 2H), 2.37(t, J = 7.3 Hz, 2H), 1.85−1.25 (m, 10H), 0.91 (t, J = 6.7 Hz, 3H); ^{13}C NMR (125 MHz, CDCl$_3$) δ ppm 194.3, 174.1, 143.2, 132.9, 129.3, 127.6, 127.1, 75.7, 73.8, 51.5, 42.3, 33.7, 31.5, 31.3, 24.3, 22.4, 21.1, 13.8; IR (neat) (v_{max}, cm^{-1}) 3270, 2952, 1741, 1658, 1413, 1297; HRMS (ESI) Found 391.1555 [M + Na]$^+$ C$_{19}$H$_{28}$O$_5$NaS requires 391.1555.

5.8.6 1-(3-Vinylthiophen-2-yl)hexan-1-one (25)

A mixture of Pd(PPh$_3$)$_4$ (22 mg, 0.0191 mmol) and LiCl (32 mg, 0.74 mmol) was dissolved in 1,4-dioxane (2.5 mL) under an atmosphere of N$_2$. Bromide **3** (100 mg, 0.382 mmol) in 1,4-dioxane (1 mL) was added slowly and followed by the addition of tributylvinylstannane (145 μL, 0.496 mmol). The reaction mixture was heated to 80 °C and stirring was continued for 12 h. The reaction mixture was filtered through a small pad of Al$_2$O$_3$ and eluted with EtOAc (100 mL). The solvent was removed in vacuo and the residue was purified using silica gel chromatography (pentane/Et$_2$O, 98:2) to afford **25** (53 mg, 70% yield) as a clear oil. TLC: R$_f$ = 0.73 (pentane/Et$_2$O, 9.5:0.5) ^{1}H NMR (500 MHz, CDCl$_3$) δ ppm

7.57 (dd, $J = 17.8$, 11.0 Hz, 1H), 7.35 (d, $J = 5.1$ Hz, 1H), 7.32 (d, $J = 5.1$ Hz, 1H), 5.71 (dd, $J = 17.8$, 1.2 Hz, 1H), 5.41 (dd, $J = 11.0$, 1.2 Hz, 1H), 2.80 (t, $J = 7.3$ Hz, 2H), 1.74–1.68 (m, 2H), 1.35–1.31 (m, 4H), 0.89 (t, $J = 7.0$, 3H) ^{13}C NMR (125 MHz, CDCl$_3$) δ ppm 194.4, 144.7, 135.9, 131.3, 129.5, 127.6, 118.6, 42.6, 31.7, 24.6, 22.7, 14.1 IR (neat) (ν_{max}, cm^{-1}) 3091.9, 2956.7, 2871.0, 1664.8, 1426.7, 1181; HRMS (ESI) Found 209.0996 [M + H]$^+$ C$_{12}$H$_{17}$OS requires 209.1000.

5.8.7 (R)-1-(3-Vinylthiophen-2-yl)hexan-1-ol ((1R)-14)

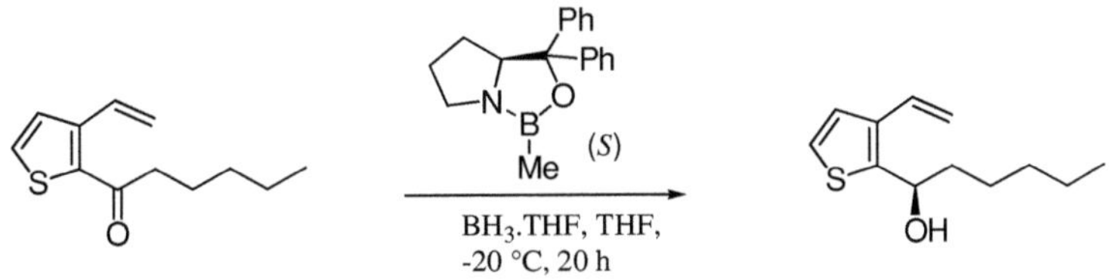

(S)-(+)-2-Methyl-CBS-oxazaborolidine (33 mg, 0.12 mmol) was dissolved in dry THF (2 mL) and this mixture was brought to $-20\ °C$ under N$_2$. BH$_3$.THF (1 M, 120 µL, 1.16 mmol) was added followed by the addition of ketone **25** (100 mg, 0.483 mmol) in THF (1 mL) and the reaction mixture was stirred at $-20\ °C$ for 24 h. MeOH (1.5 mL) was added slowly and H$_2$ was given off. The residue was dry loaded onto a silica gel column for purification (pentane/CH$_2$Cl$_2$, 4:1 then 1:1) to afford (1R)-**14** (66 mg, 65% yield) as a colourless oil. $ee = 94\%$ as determined by chiral HPLC, Chiracel$^®$ OD column: 99:1 hexane/2-propanol, 1.0 mL/min, $t_R = 23.3$ min   for (S), $t_R = 26.6$ min   for (R).   TLC: $R_f = 0.31$(pentane/CH$_2$Cl$_2$, 4:1), $[\alpha]_D^{20}$ +8.3 ($c = 0.91$, CHCl$_3$). ^{1}H NMR (500 MHz, CDCl$_3$) δ ppm 7.16 (br. s, 2H), 6.78 (dd, J = 17.4, 10.9 Hz, 1H), 5.56 (dd, J = 17.4, 1.1 Hz, 1H), 5.24 (dd, J = 17.4, 1.1 Hz, 1H), 5.09 (m,1H), 1.95 (br. d, 1H), 1.93–1.86 (m, 1H), 1,81–1.74 (m, 1H), 1,49-1.29 (m,6H), 0.88 (t, J = 6.87 Hz, 3H); ^{13}C NMR (125 MHz, CDCl$_3$) δ ppm 144.7, 135.8, 128.9, 125.2, 123.7, 114.5, 68.2, 39.4, 31.6, 25.6, 22.6, 14.0; IR (neat) (ν_{max}, cm^{-1}) 3358, 2954, 2929, 2857, 1459, 1242; HRMS (ESI) Found 211.1157 [M + H]$^+$ C$_{12}$H$_{19}$OS requires 211.1157.

5.8.8 Methyl 4-((4S,5R)-5-((E)-2-(2-((R)-1-hydroxyhexyl)thiophen-3-yl)vinyl)-2,2 dimethyl-1,3-dioxolan-4-yl)butanoate((1R)-13)

Grubbs (II) catalyst (18.5 mg, 0.0219 mmol) was dissolved in dry CH$_2$Cl$_2$ (2 mL) under N$_2$ to which alcohol (1R)-**14** (46 mg, 0.219 mmol) in CH$_2$Cl$_2$ (1 mL) and olefin **26** (60 mg, 0.26 mmol) in CH$_2$Cl$_2$ (1 mL) was added. The reaction mixture was stirred at 40 °C for 96 h. The mixture was dry loaded onto a silica column for purification (pentane/EtOAc, 9:1 then 8:2) to afford (1R)-**13** (29 mg, 32% yield) as a yellow oil. TLC: R$_f$ = 0.25 (pentane/EtOAc, 4:1) [α]$_D^{20}$ +1.8 (c = 1.2, CHCl$_3$); ^{1}H NMR (500 MHz, CDCl$_3$) δ ppm 7.16 (d, J = 5.1 Hz, 1H), 7.13 (d, J = 5.1 Hz, 1H), 6.70 (d, J = 15.6 Hz, 1H), 5.98 (dd, J = 15.6, 7.9 Hz, 1H), 5.07 (m, 1H), 4.65 (t, J = 6.9 Hz, 1H), 4.18 (m, 1H), 3.63 (s, 3H), 2.34 (m, 2H), 2.38–2.30 (m, 2H), 2.02 (br. s, 1H), 1.51(s, 3H), 1.39 (s, 3H), 1.92–1.43 (m, 6H), 1.33–1.29 (m, 4H), 0.88 (app. s, 3H); ^{13}C NMR (125 MHz, CDCl$_3$) δ ppm 173.8, 145.0, 134.5, 126.1, 125.7, 125.5, 123.8, 108.3, 79.6, 78.3, 68.2, 51.5, 39.3, 33.8, 31.6, 30.1, 28.3, 25.7, 25.6, 22.6, 21.7, 14.0. IR (neat) (v_{max}, cm^{-1}) 3445, 2985, 2930, 1737, 1456, 1245; (ESI) Found 433.2028 [M + Na]$^+$ C$_{22}$H$_{34}$O$_5$NaS requires 433.2025.

5.8.9 Methyl 4-((4S,5R)-2,2-dimethyl-5-vinyl-1,3-dioxolan-4-yl)butanoate (26)

Diol **15** (100 mg, 0.531 mmol) was dissolved in dichloromethane (6 mL) to which 2,2-dimethoxypropane (0.097 mL, 0.796 mmol) and p-TSA (10.2 mg, 0.053 mmol) was added under an atmosphere of N$_2$. The reaction mixture was allowed to stir at ambient temperature for 24 h. The solvent was removed and the residue was purified by silica gel chromatography (pentane/ethyl acetate, 6:1) to afford **26** as a colourless oil (93 mg, 75% yield). TLC: $R_f = 0.62$ (pentane/EtOAc, 9:1); $[\alpha]_D^{20}$ −7.2 (c = 1.0, CHCl$_3$); ^{1}H NMR (500 MHz, CDCl$_3$) δ ppm 5.78 (ddd, J = 17.1, 10.3, 7.8 Hz, 1H), 5.31–5.20 (m, 2H), 4.48 (t, J = 7.2 Hz, 1H), 4.14–4.10 (m, 1H), 3.65 (s, 3H), 2.33 (t, J = 7.2 Hz, 2H), 1.81–1.61 (m, 2H), 1.53–1.38 (m, 2H), 1.46 (s, 3H), 1.35 (s, 3H) ppm; ^{13}C NMR (100 MHz, CDCl$_3$) δ ppm 173.8, 134.3, 118.3, 108.2, 79.7, 77.9, 51.5, 33.8, 29.9, 28.2, 25.6, 21.7 ppm; IR (neat) (ν_{max}, cm^{-1}) 2987, 2952, 1739, 1380, 1216; HRMS (ESI) Found 251.1265 [M + Na]$^+$ C$_{12}$H$_{20}$O$_4$Na requires 251.1259.

5.8.10 Cytokine Production by J774 Macrophages

The murine J774 macrophages (European Collection of Cell Cultures, UK) were maintained in suspension of RPMI 1,640 supplemented with 2 mmol/l glutamine, 100 IU/mL penicillin, 100 µg/mL streptomycin, and 10% fetal calf serum (Life Technologies Inc., Grand Island, NY). Cells were seeded at 1×10^6 cells per ml for experiments and exposed to lipopolysaccharide (LPS) at a concentration of 100 ng/mL for 24 h at 37 °C in 5% CO$_2$. **10** and (1R)-**13** were added to the cells 1 h before addition of LPS, at a concentration of 1 nM, 1 µM and 10 µM. After 24 h the supernatants were collected for cytokine analysis. IL-12p40, TNF-α, IL-1β, IL-6, MCP, MIP-1 and MIP-2 concentrations in cell culture supernatants were quantified by commercial DuoSet ELISA kits (R&D Systems), according to the manufacturer's instructions.

References

1. O' Sullivan TP, Vallin KSA, Shah STA, Fakhry J, Maderna P, Scannell M, Sampaio ALF, Perretti M, Godson C, Guiry PJ (2007) J Med Chem 50:5894
2. Petasis NA, Akritopoulou-Zanze I, Fokin VV, Bernasconi G, Keledjian R, Yang R, Uddin J, Nagulapalli KC, Serhan CN (2005) Prostaglandins. Leukot Essent Fat Acids 73:301
3. Sun Y-P, Tjonahen E, Keledjian R, Zhu M, Yang R, Recchiuti A, Pillai PS, Petasis NA, Serhan CN (2009) Prostaglandins Leukot Essent Fat Acids 81:357
4. Duffy CD, Maderna P, McCarthy C, Loscher CE, Godson C, Guiry PJ (2010) Chem Med Chem 5:517
5. Chen X, Wang W (2003) Annu Rep Med Chem 38:333
6. Patani GA, LaVoie EJ (1996) Chem Rev 96:3147
7. Siebert CD (2004) Chem Unserer Zeit 38:320
8. Tai C-L, Hung M-S, Pawar VD, Tseng S-L, Song J-S, Hsieh W-P, Chiu H-H, Wu H-C, Hsieh M-T, Kuo C-W, Hsieh C-C, Tsao J-P, Chao Y-S, Shia K-S (2008) Org Biomol Chem 6:447
9. Oberdorf C, Schepmann D, Vela JM, Diaz JL, Holenz J, Wünsch B (2008) J Med Chem 51:6531
10. Chao J, Taveras AG, Aki CJ (2009) Tetrahedron Lett 50:5005
11. Bitter I, Dossenbach MRK, Brook S, Feldman PD, Metcalfe S, Gagiano CA, Füredi J, Bartko G, Janka Z, Banki CM, Kovacs G, Breier A (2004) Prog Neuropsychopharmacol Biol Psychiatry 28:173

12. Cohen H, Loewenthal U, Matar M, Kotler M (2001) Br J Psychiatry 179:167
13. Alvir JMJ, Lieberman JA, Safferman AZ, Schwimmer JL, Schaaf JA (1993) N Engl J Med 329:162
14. http://drugtopics.modernmedicine.com
15. Fuller LS, Iddon B, Smith KA (**1997**) J Chem Soc Perkin Trans 1 22:3465
16. Corey EJ, Suggs JW (1975) Tetrahedron Lett 16:2647
17. Agarwal S, Tiwari HP, Sharma JP (1990) Tetrahedron 46:4417
18. Srebnik M, Ramachandran P, Brown HC (1988) J Org Chem 53:2916
19. Vaino AR, Szarek WA (1996) Chem Commun 2351
20. Corey EJ, Venkateswarlu A (1972) J Am Chem Soc 94:6190
21. Kende AS, Liu K, Kaldor I, Dorey G, Koch K (1995) J Am Chem Soc 117:8258
22. Prakash C, Saleh S, Blair IA (1989) Tetrahedron Lett 30:19
23. Scholl M, Ding S, Lee CW, Grubbs RH (1999) Org Lett 1:953
24. Grubbs RH, Chang S (1998) Tetrahedron 54:4413
25. Trnka TM, Grubbs RH (2000) Acc Chem Res 34:18
26. Sheddan NA, Mulzer J (2006) Org Lett 8:3101
27. Sheddan NA, Arion VB, Mulzer J (2006) Tetrahedron Lett 47:6689
28. Kawai T, Shida Y, Yoshida H, Abe J, Iyoda T (2002) J Mol Catal A Chem 190:33
29. Milstein D, Stille JK (1978) J Am Chem Soc 100:3636
30. Jyotirmayee D, Stellios A, Cossy J (2007) Adv Synth Catal 349:152
31. Hoffman TJ, Rigby JH, Arseniyadis S, Cossy J (2008) J Org Chem 73:2400
32. Corey EJ, Bakshi RK, Shibata S (1987) J Am Chem Soc 109:5551
33. Wu S-H, Liao PY, Dong L, Chen ZQ (2008) Inflamm Res 57:430
34. Decker Y, McBean G, Godson C (2009) Am J Physiol Cell Physiol 296:C1420
35. Aliberti J, Serhan C, Sher A (2002) J Exp Med 196:1253
36. Sodin-Semrl S, Taddeo B, Tseng D, Varga J, Fiore S (2000) J Immunol 164:2660

Chapter 6
Towards the Synthesis of Various Heteroaromatic Lipoxin A₄ Analogues

6.1 Introduction

Our research group has successfully demonstrated that the introduction of benzene, pyridine and thiophene rings into the core Lipoxin structure has contributed to an enhancement of the biological profile of this class of eicosanoid [1, 2]. There is an on-going effort in the pharmaceutical industry to design and synthesise new drugs to combat existing inflammatory disorders. These novel stable LXA₄ analogues possess the ability to aid the inflammation process and are therefore showing potential as therapeutic agents. Taking these recent advances into account, we have designed a series of novel heteroaryl LXA₄ analogues, Fig. 6.1, with a view to further increasing the biological potency of these anti-inflammatory agents.

We have previously established an efficient route for the synthesis of the Heck coupling partner, Chap. 3, which forms the top chain of the Lipoxin molecule. This chain is common in all three heteraryl LXA₄ analogues, Fig. 6.1. The key synthetic transformation for the preparation of these analogues relies on the palladium-catalysed Heck reaction to furnish the required *trans* olefin. In addition to this, an organometalic carbon–carbon bond forming reaction will be employed to introduce the lower chain of the molecule.

6.2 Towards the Synthesis of 6-Methyl Pyridine LXA₄ 1

We have previously enhanced the pharmacological profile of our aromatic LXA₄ analogues by the addition of a heteroatom into the Lipoxin structure. In an extension of this work, we sought to increase the bioactivity by probing the steric and electronic properties of these heterocycles. We focused our efforts on replacing one of the aromatic protons with a methyl group and studying the

C. Duffy, *Heteroaromatic Lipoxin A₄ Analogues*, Springer Theses,
DOI: 10.1007/978-3-642-24632-6_6, © Springer-Verlag Berlin Heidelberg 2012

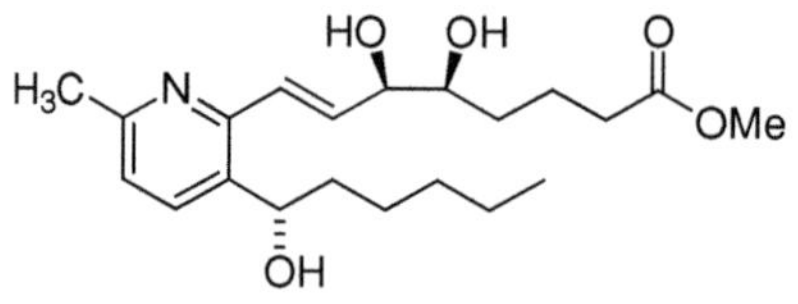

6-Methyl pyridine LXA_4

1

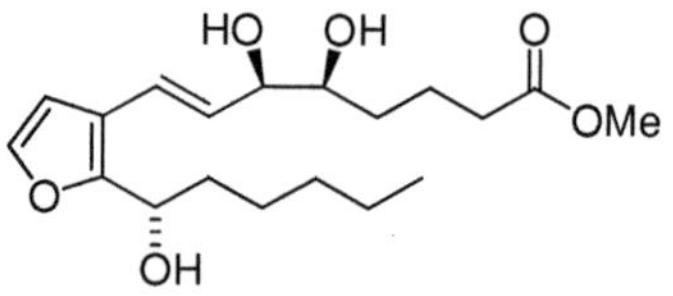

Furan LXA_4

2

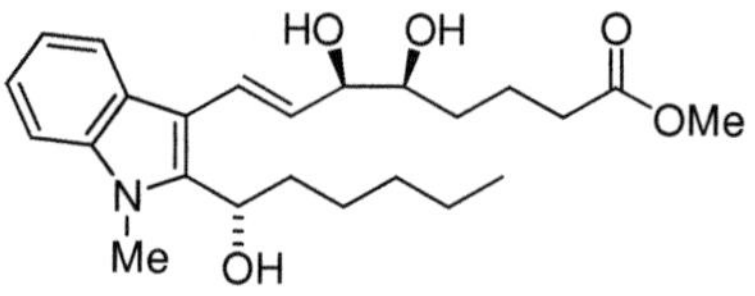

Indole LXA_4

3

Fig. 6.1 A series of heteraryl LXA_4 analogues

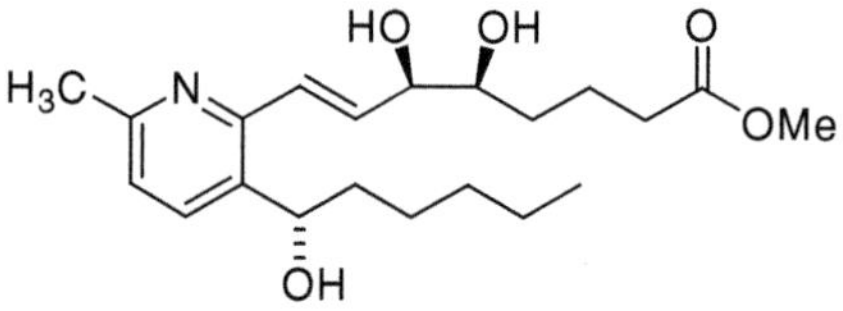

6-Methyl pyridine LXA $_4$

Fig. 6.2 Proposed 6-methyl pyridine LXA_4 analogue

effect that this substitution has on the biological potency of this compound, Fig. 6.2.

The synthetic route for the preparation of this analogue relies on the palladium-catalysed Heck reaction of bromide **4** and olefin **5**, Scheme 6.1.

The first step in the synthesis required the one-pot bromination and esterification of commercially available acid **6**, Scheme 6.2 [3].

The ester **7** was successfully prepared by the addition of phosphoryl oxybromide in chlorobenzene and pyridine at 145 °C for 3 h in 64% yield. The formation of the product was confirmed by the appearance of a singlet at 3.94 ppm integrating for three protons in the ^{1}H NMR spectrum and the pair of doublets in the aromatic region at 7.18 and 7.99 ppm, Fig. 6.3.

6-Methyl pyridine LXA$_4$

1

Scheme 6.1 Reterosynthetic analysis of 6-methyl pyridine LXA$_4$ **1**

Scheme 6.2 One-pot bromination and esterification of acid **6** [3]

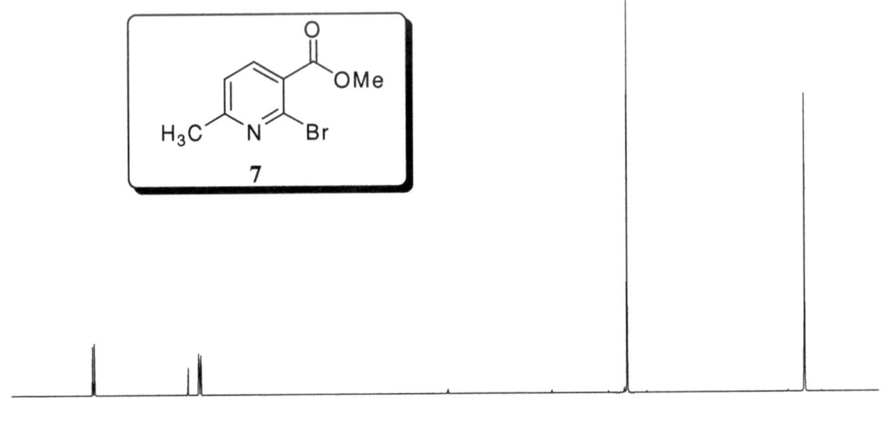

Fig. 6.3 500 MHz ^{1}H NMR spectrum of ester **7**

Ester **7** was subsequently converted into its corresponding acid **8** by the addition of LiOH in a mixture of MeOH and water for 1 h in 97% yield, Scheme 6.3.

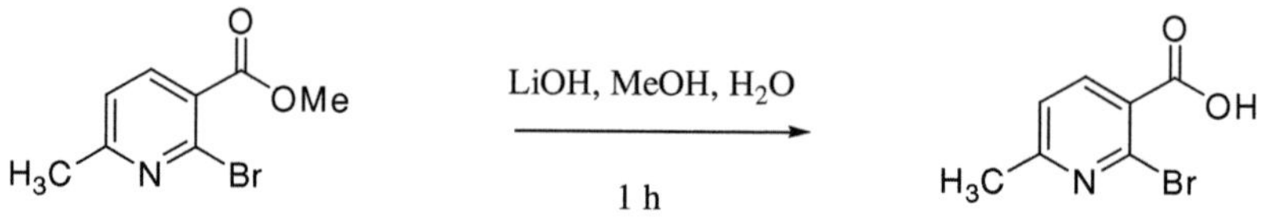

Scheme 6.3 Synthesis of acid **8**

The next step in the synthesis required the activation of the acid **8** using thionyl chloride and subsequent reaction of this acid chloride with the Grignard derivative of 1-bromopentane, Scheme 6.4. The additional use of bis[2-(N,N-dimethyl-amino)ethyl]ether in this reaction moderated the reactivity of the Grignard reagents and prevents any double addition occurring [4]. Unexpectedly, during this reaction, the aryl bromide exchanged to give the corresponding aryl chloride **9** in 27% yield.

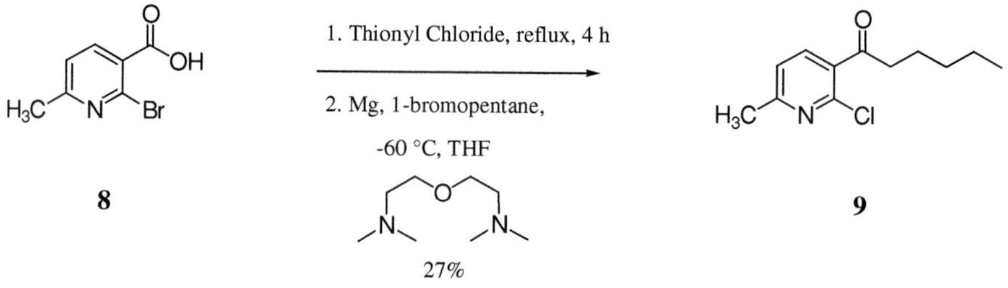

Scheme 6.4 Unexpected aryl chloride **9** formation

The formation of this product **9** was confirmed by High Resolution Mass Spectrometry with the isotope pattern characteristic of chloride. The ^{1}H NMR spectrum revealed that the newly formed carbon–carbon bond had formed with the presence of a triplet at 2.97 ppm integrating for 2 protons, Fig. 6.4.

It is well established that Heck reaction conditions used for the coupling of aryl bromides, iodides and triflates with olefins are not best suited to coupling with aryl chlorides [5]. This low reactivity is believed to be caused by the strength to the C–Cl bond and its inability to undergo oxidative addition during the

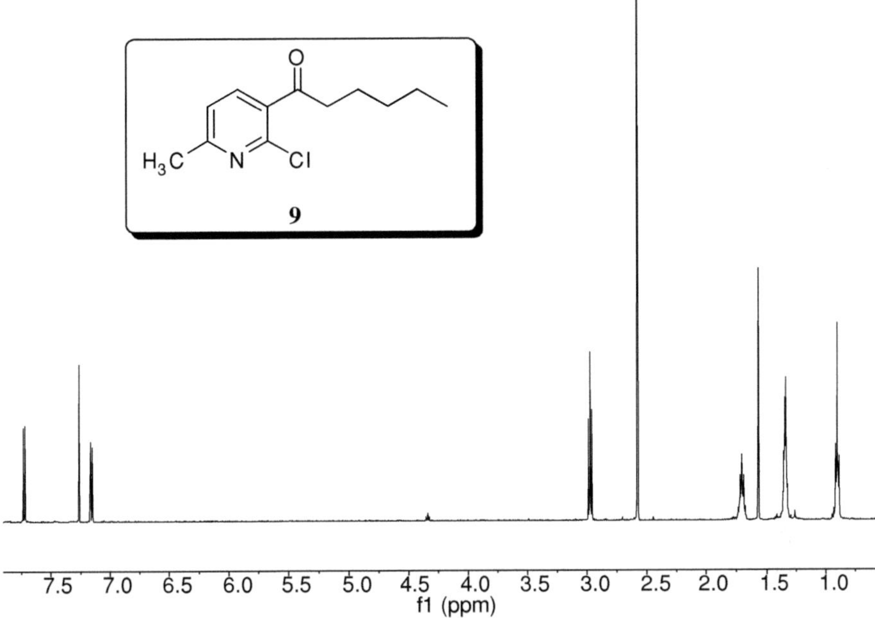

Fig. 6.4 500 MHz ¹H NMR spectrum of ketone **9**

Table 6.1 Attempted Heck reactions for aryl chloride **12**

Pd source	Ligand	Base	Solvent	Temp. (°C)	Time	% Yield
(C₃H₅)₂Pd₂Cl₂ (5 mol%)	(o-tolyl)₃P	NaOAc	Toluene : DMA (3:1)	115	12 h	No reaction
Pd(OAc)₂ (20 mol%)	(o-tolyl)₃P	PMP	Acetonitile	100	7 days	No reaction
Pd(OAc)₂ (10 mol%)	(o-tolyl)₃P	Bu₃N	Bu₃N	120	24 h	No reaction

Heck reaction [6]. With this in mind, we were not surprised to find that a variety of Heck reaction conditions failed to produce the required *trans* product **10** and only starting materials were recovered, Table 6.1.

In conclusion, the synthesis of this analogue **1** remains a challenge and future efforts in its preparation will require the synthesis of the key intermediate aryl bromide **4** without the formation of any aryl chloride **9**. We are confident that the Heck reaction would be successful if the aryl bromide was used, as similar substrates are reported in the literature for this reaction [2, 3]. The synthesis will provide key information to our Structure Activity Relationship Study and potentially inspire the synthesis of diverse Lipoxin analogues whereby the bioactivity can be tuned by various substitutions on the ring.

6.3 Towards the Synthesis of Furan LXA$_4$ 2

We have previously demonstrated that bioactivity in the stable LXA$_4$ analogues can be retained following the replacement of benzene with thiophene, Chap. 5. These biologically active analogues were capable of stimulating or hindering the production of key cytokines. We therefore attempted to synthesis a novel furan-containing LXA$_4$ analogue **2**, Fig. 6.5.

Substitution of benzene for furan is another classical example of successful bioisosteric replacement in medicinal chemistry. The furan moiety is well represented in current drugs on the market, for example the histamine H$_2$-receptor antagonist Zantac was one of the first "blockbuster drugs" for Glaxo with annual sales over \$1 billion, Fig. 6.6.

Furan analogue **2** was designed with the intention of further probing the bioactivity of these stable analogues. The aim was to compare the previously prepared stable five-membered heteroaryl LXA$_4$ analogues, described in Chap. 5. The proposed synthetic route incorporates a palladium-catalysed Heck reaction and a regioselective α-lithiation of commercially available 3-bromofuran **11**. The efficient deprotonation of 3-bromofuran reported in the late 1970s allows for the formation of 2,3-substituted furans in good yields [7]. Treating 3-bromofuran **11** with freshly prepared lithium diisopropylamide afforded the α-lithiated intermediate which was quenched with hexanal **12** at −78 °C in THF to give alcohol **13** in 63% yield, Scheme 6.5.

Scheme 6.5 Regioselective α-lithiated of 3-bromofuran **13**

Evidence for the formation of **13** was observed in the ^{1}H NMR spectrum with the presence of a multiplet at 4.79 ppm integrating for one proton, corresponding to the CH directly attached to the hydroxyl group, Fig. 6.7. The presence of this hydroxy group was also observed in IR spectrum as a broad stretch at 3,347 cm^{-1}.

The next step in the synthesis required the preparation of ketone **14** via the oxidation of alcohol **13**. Furan-containing compounds are prone to decomposition

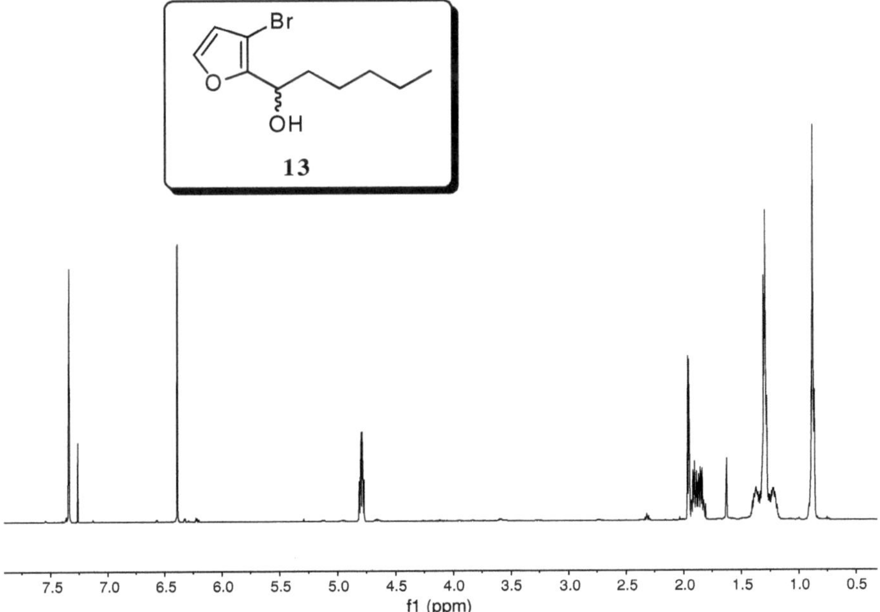

Furan LXA$_4$

2

Fig. 6.5 Proposed furan-containing LXA$_4$ analogue **2**

Fig. 6.6 Furan-containing drug Zantac

Fig. 6.7 500 MHz ^{1}H NMR of alcohol **13**

in acidic environments [8], therefore, Swern oxidation conditions were employed
for this transformation as this is a well established procedure for the oxidation of
secondary alcohols, Scheme 6.6 [9]. Treating alcohol **13** with oxalyl chloride in

Table 6.2 Attempted Heck reactions conditions

Pd source	Ligand	Base	Solvent	Temp. (°C)	Time	% Yield
Pd(OAc)$_2$ (10 mol%)	(o-tolyl)$_3$P	Bu$_3$N	Bu$_3$N	120	24 h	No reaction
Pd(OAc$_2$ (10 mol%)	(o-tolyl)$_3$P	Bu$_3$N	Bu$_3$N + AgOAc	120	24 h	No reaction
(C3H5)$_2$Pd$_2$Cl$_2$ (5 mol%)	(o-tolyl)$_3$P	NaOAc	Toluene:DMA (3:1)	115	12 h	No reaction
Pd(OAc)$_2$ (20 mol%)	(o-tolyl$_3$P	PMP	Acetonitrile	100	7 days	No reaction
Pd(PPh$_3$)$_4$ (5 mol%)	N/A	PMP	Acetonitrile	100	72 h	No reaction

DMSO followed by the addition of triethylamine afforded ketone **14** in a high 82% yield.

Scheme 6.6 Swern oxidation of alcohol **13**

The ^{1}H NMR spectrum of ketone **13** confirmed the formation of the product as a triplet was observed at 2.88 ppm integrating for two protons corresponding to the CH$_2$ directly beside the newly formed carbonyl. A signal at 189.5 ppm in the ^{13}C NMR spectrum, along with a sharp stretch at 1,681 cm^{-1} in the IR spectrum, confirmed the oxidation had been successful.

This ketone **14** was subsequently used in several Heck reactions with olefin **5** in an attempt to produce the required *trans* olefin **15**, Table 6.2. Unfortunately, all of the reactions carried out with this ketone failed to give the desired product **15**. Both the olefin and ketone starting materials were recovered which indicated that oxidative addition had not occurred.

This severe lack of reactivity towards Heck reaction conditions directed us towards an alternative approach for the construction of the *trans* olefin. In light of the relative success of the Grubbs' cross metathesis strategy, detailed in Chap. 5, for the synthesis of thiophene-containing LXA$_4$, we envisaged a similar synthetic pathway for our furan analogues, Scheme 6.7 [10–12].

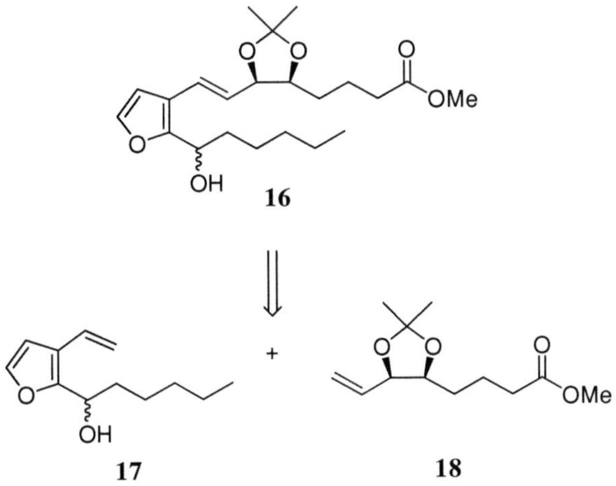

Scheme 6.7 Proposed Grubbs' cross metathesis strategy

The synthesis of vinyl alcohol **17** was accomplished through a palladium-catalysed Stille coupling reaction with bromide **14**, followed by reduction, of ketone **19**, using sodium borohyride, Scheme 6.8 [13, 14].

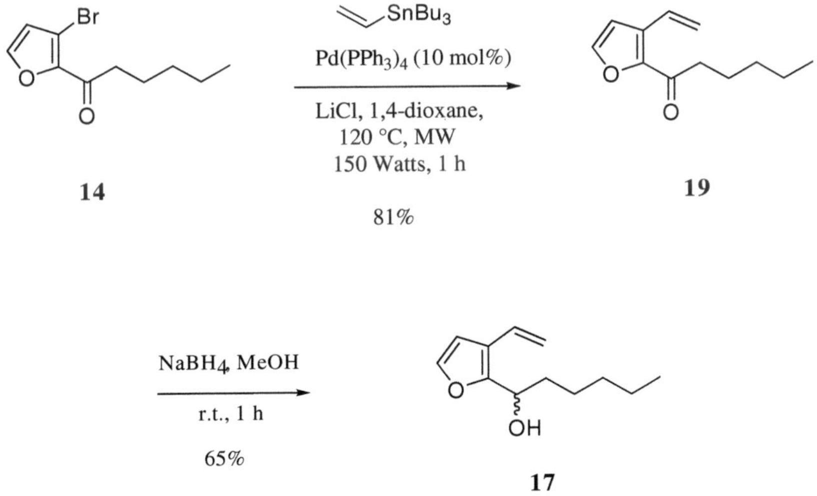

Scheme 6.8 Synthesis of vinyl alcohol **17**

This vinyl alcohol **17** was reacted with olefin **18** via a cross metathesis reaction using Grubbs' 2nd generation catalyst. However, under these conditions, no reaction took place, Scheme 6.9.

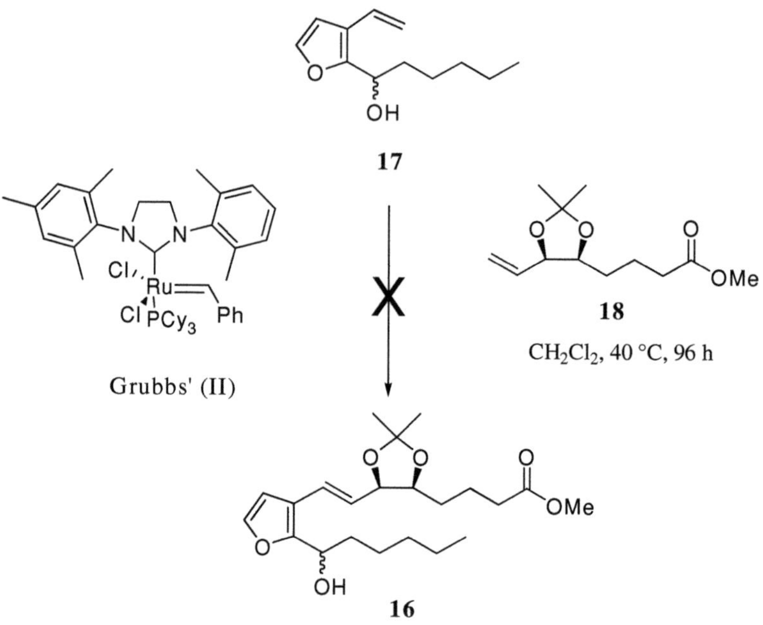

Scheme 6.9 Attempted Grubbs' cross metathesis

Future efforts will focus on designing a new strategy for the preparation of furan-containing LXA$_4$ analogue **2**. Having realised that the Grubbs' cross metathesis coupling is not a feasible route to this analogue, future work will aim to develop a new synthesis, overcoming this problem in order to provide the desired heteroaryl analogue. We would then hope to assess the biological significance of replacing the active benzene moiety with a furan ring.

6.4 Towards the Synthesis of Indole LXA$_4$ 3

We have recently shown how five- and six-member heterocycles can enhance the bioactivity of stable Lipoxin analogues. We therefore designed and attempted the synthesis of a novel indole-containing LXA$_4$ analogue **3**, Fig. 6.8.

Indole is an extremely common fused heterocycle found in a diverse range of natural products. Along with its occurence in natural products isolated from plants and fungi, the indole moiety is found in the essential amino acid Tryptophan and neurotransmitter Serotonin, Fig. 6.9 [15].

This natural occurrence and bioactivity has inspired the pharmaceutical industry to design and synthesise many indole-containing drugs to treat a broad range of illnesses including schizophrenia, asthma and Parkinson's disease [15, 16].

The first step in the synthesis proceeds with the functionalisation of commercially available indole **20**, to give the 2,3-dibrominated indole product **21**,

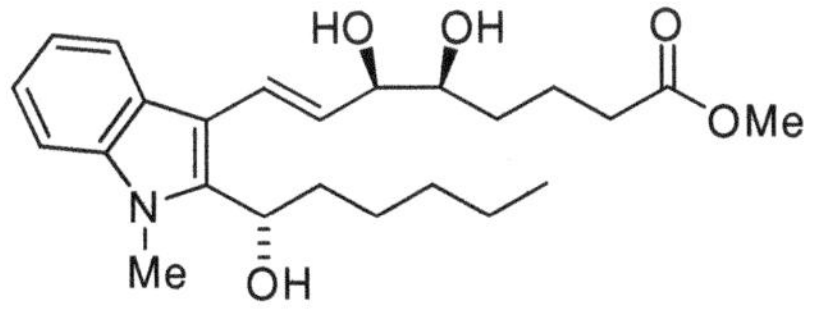

Indole LXA$_4$

3

Fig. 6.8 Design of indole-containing LXA$_4$ analogue **3**

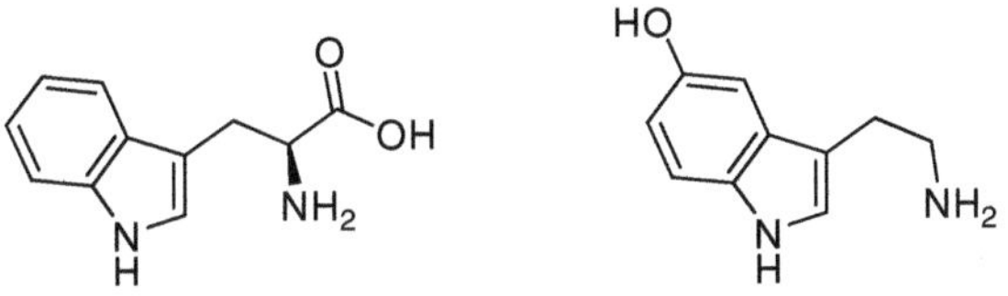

Tryptophan   Serotonin

Fig. 6.9 Naturally occurring indole-containing compounds

Scheme 6.10. This was accomplished with the use of Bergman's efficient synthesis for the formation of 2-halo-indoles [19], followed directly by a one-pot bromination–methylation reaction [17–19].

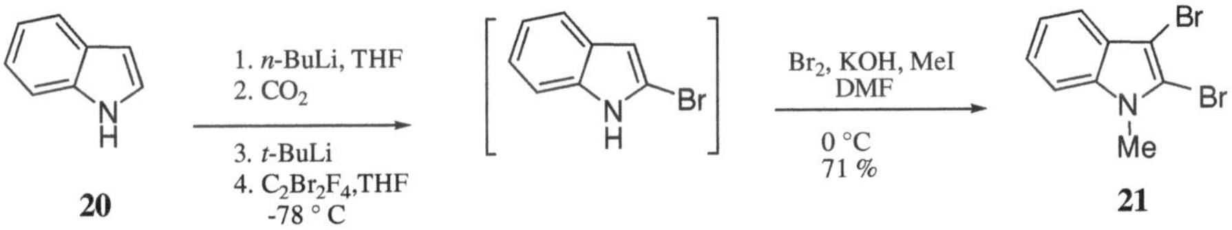

Scheme 6.10 Prepartion of dibrominated indole intermediate **21**

Analysis of the ^{1}H NMR and ^{13}C NMR spectra of **21** confirmed that the methylation and di-bromination reactions had successfully proceeded in one-pot. Careful study of the aromatic region in the ^{1}H NMR spectrum, Fig. 6.10, revealed two doublets and two triplets accounting for the four aromatic protons. This was also accompanied by a singlet at 3.82 ppm integrating for three protons corresponding to the newly formed *N*-methyl group.

The construction of the lower chain was the next synthetic challenge in the synthesis of this analogue. Gribble and co-workers have demonstrated that a lithium/halogen exchange reaction of **21** occurs smoothly by the addition of *tert*-butyllithium in THF. The authors also show the versatility of this exchange reaction by trapping the lithio-intermediate **22** with a variety of electrophiles, Scheme 6.11 [20, 21].

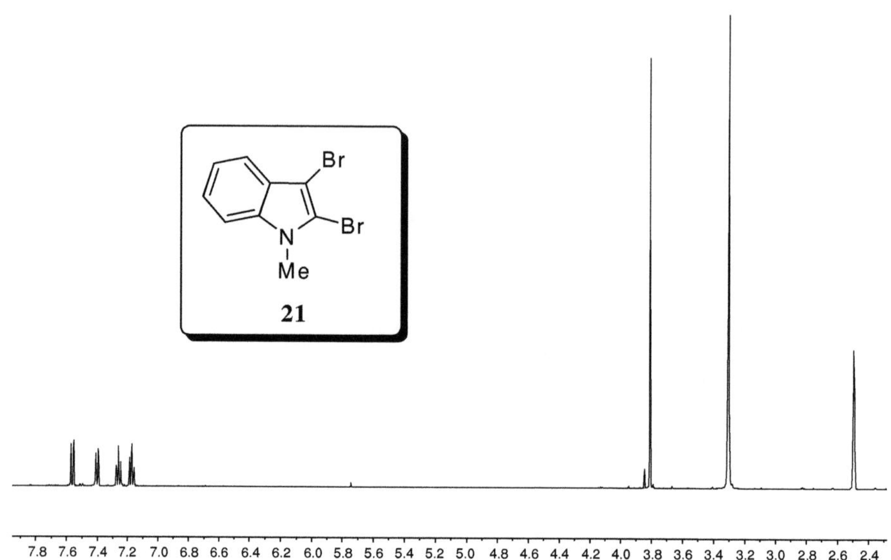

Fig. 6.10 500 MHz ^{1}H NMR of **21** in D$_6$-DMSO

Scheme 6.11 Lithium/halogen exchange followed by quenching with various electrophiles [20]

We therefore carried out this reaction by quenching the lithio-intermediate **22** with hexanal, Scheme 6.12, to afford the desired alcohol **23** in 53% yield. This was followed by a Swern oxidation to afford the corresponding ketone **24** for the palladium-catalysed Heck reaction.

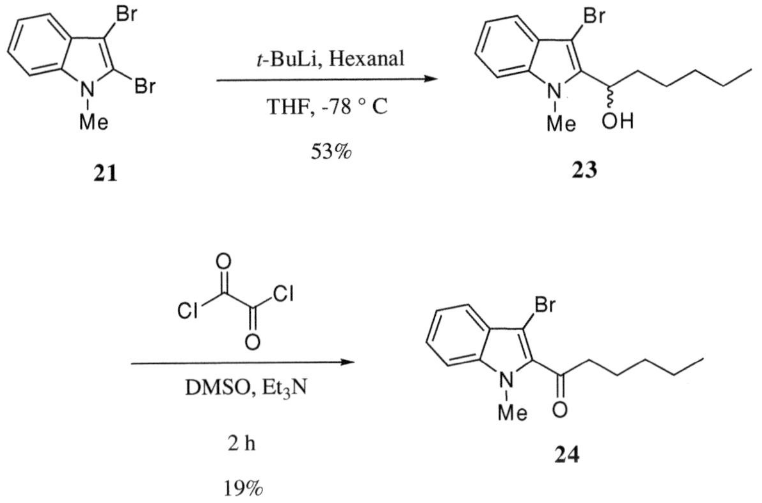

Scheme 6.12 Preparation of ketone **24**

Unfortunately, the Heck coupling of ketone **24** with olefin **5** failed to produce the desired *trans* product, Scheme 6.13. No reaction took place which indicated that oxidative addition did not occur in this case.

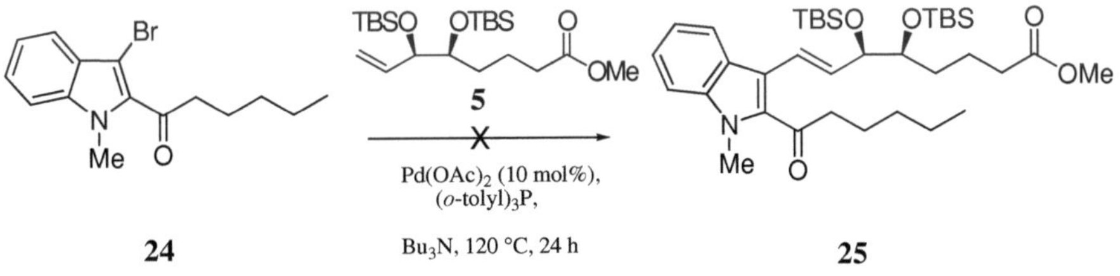

Scheme 6.13 Attempted Heck reaction

6.5 Conclusion

In summary, we have successfully synthesised three new heteroaryl intermediates in our route towards the synthesis of interesting novel LXA$_4$ analogues. Problems in the synthesis of all three analogues arose during the cross coupling step which aimed to incorporate the *trans* olefin. The fact that these three heteroaryl analogues contain biologically important motifs, namely pyridine, furan and indole moieties, inspired us to reconsider the current synthetic route. As with the other heteroaryl analogues, these new analogues will be evaluated by their ability to promote the clearance of PMNs and their effect on the production of key cytokines.

6.6 Experimental

6.6.1 Methyl 2-Bromo-6-Methylnicotinate (7)

Phosphoryl tribromide (12.9 g, 45.1 mmol) was added in small portions to a solution of acid **6** (3 g, 19.6 mmol), pyridine (1.53 mL, 19.6 mmol) and chlorobenzene (60 mL) at room temperature under an atmosphere of nitrogen. The mixture was heated to reflux for 3 h and concentrated under vacuum. Cold methanol (20 mL) was added slowly and the solution was stirred for 1 h and concentrated under vacuum. CH_2Cl_2 (30 mL) was added followed by cold H_2O (30 mL). The pH of the solution was adjusted to 8 using K_2CO_3. This was extracted using CH_2Cl_2 (3 × 60 mL). The organic layers were washed with 10% Na_2CO_3 (60 mL) and saturated ammonium chloride (60 mL) and dried over $MgSO_4$. The residue was purified using silica gel chromatography (CH_2Cl_2/pentane, 4:1) to afford **7** (2.9 g, 64% yield) as an orange oil. (Lit.[3]) TLC: $R_f = 0.29$ (CH_2Cl_2/pentane acetate, 4:1); ^{1}H NMR (500 MHz, CDCl$_3$) δ 7.99 (d, $J = 7.8$ Hz, 1H), 7.18 (d, $J = 7.8$ Hz, 1H), 3.94 (s, 3H), 2.59 (s, 3H), ppm; ^{13}C NMR (75 MHz, CDCl$_3$) δ 165.6, 162.9, 140.3, 140.2, 126.4, 122.1, 52.9, 24.5 ppm; IR (neat) (ν_{max}, cm^{-1}) 1732, 1587, 1431, 1344, 1277, 1141, 1049; HRMS (ESI) Found 229.9819 [M + H]$^+$, $C_8H_9BrNO_2$ requires 229.9817.

6.6.2 1-(2-Bromo-6-Methylpyridin-3-yl)Hexan-1-One (9)

Ester **7** (1.2 g, 5.2 mmol) was dissolved in MeOH (10 mL) and H_2O (1 mL) to which LiOH (254 mg, 10.5 mmol) was added. The reaction mixture was stirred for

1 h and the solvent was removed and the resulting mixture was purified by passing through a very short silica gel column (CH$_2$Cl$_2$/MeOH, 9:1) to afford a white solid. This material was sufficiently pure to be carried on to the next step without the need for any further purification. TLC: R$_f$ = 0.20 (CH$_2$Cl$_2$/MeOH, 9:1); ^{1}H NMR (500 MHz, CD$_3$OD) δ 7.56 (d, J = 7.5 Hz, 1H), 7.12 (d, J = 7.5 Hz, 1H), 2.36 (s, 3H), ppm; HRMS (ESI) Found 213.9513 [M − H]$^-$, C$_7$H$_5$BrNO$_2$ requires 213.9504. This acid was dissolved in thionyl chloride (6 mL, 1.15 mol) and heated to reflux under an atmosphere on nitrogen for 4 h. The excess thionyl chloride was removed under vacuum to give the corresponding acid chloride which was used in the next step without any further purification. The Grignard derivative of 1-bromopentane (770 mg, 5.1 mmol) was prepared by the addition of the bromide to preactivated Mg turnings (122 mg, 5.1 mmol) under nitrogen in THF (6 mL) and refluxed for 45 min. This Grignard derivative was added to a flask containing dimethyl aminoethyl ether (971 mg, 5.1 mmol) in THF (5 mL) at 0 °C and stirring was continued for 15 min. This solution was added to the acid chloride in THF (2 mL) over 15 min at −60 °C and stirred under nitrogen for 15 min. This mixture was quenched with saturated ammonium chloride (5 mL) and extracted using EtOAc (3 × 25 mL) and dried over MgSO$_4$. The remaining residue was dry loaded onto a silica gel column for purification (pentane/ethyl acetate, 9:1) to afford **9** (150 mg, 27% yield) as an orange oil. TLC: R$_f$ = 0.38 (pentane/ethyl acetate, 9.5:0.5) ^{1}H NMR (500 MHz, CDCl$_3$) δ 7.73 (d, J = 7.7 Hz, 1H), 7.16 (d, J = 7.7 Hz, 1H), 2.97 (t, J = 7.7 Hz, 2H) 2.58 (s, 3H), 2.59 (s, 3H), 1.74–1.68 (m, 2H), 1.36–1.32 (m, 3H), 0.90 (m, 3H) ppm. ^{13}C NMR (125 MHz, CDCl$_3$) δ 202.0, 161.4, 146.7, 138.5, 132.7, 122.0, 42.7, 31.3, 24.2, 23.9, 22.4, 13.9; IR (neat) (ν_{max}, cm^{-1}) 2957, 2930, 2871, 1699, 1587, 1440, 1346; HRMS (ESI) Found 226.0644 [M + H]$^+$, C$_{12}$H$_{17}$NOCl requires 226.0999.

6.6.3 1-(3-Bromofuran-2-yl)Hexan-1-ol (13)

3-Bromofuran **11** (1.5 mL, 17.0 mmol) was added dropwise to a stirred solution of lithium diisopropylamide prepared by addition of *n*-butyllithium (7.4 mL, 2.5 M in hexane, 18.7 mmol) to diisopropylamine (2.4 mL, 17.0 mmol) in tetrahydrofuran (70 mL) at −78 °C and the resulting mixture was stirred for a further 2 h at this temperature. Hexanal **12** (4.1 mL, 34.0 mmol) in THF (5 mL) was added and the mixture was stirred for 1 h at −78 °C and warmed to room temperature.

The mixture was quenched with saturated aqueous ammonium chloride (50 mL) and extracted with diethyl ether (3 × 50 mL). The combined extracts were washed with brine (50 mL) and dried over Na$_2$SO$_4$. The solvent was evaporated and the resulting oil was purified by silica gel column chromatography (pentane/ethyl acetate, 9.5:0.5) to afford **13** (2.6 g, 63%) as a colourless oil; TLC: R$_f$ = 0.26 (pentane/ethyl acetate, 9.5:0.5); ^{1}H NMR (500 MHz, CDCl$_3$) δ 7.34 (d, J = 1.9 Hz, 1H), 6.39 (d, J = 1.9 Hz, 1H), 4.79 (m, 1H), 1.95 (br. s, 1H), 1.92–1.84 (m, 2H), 1.40–1.18 (m, 6H), 0.87 (m, 3H) ppm; ^{13}C NMR (75 MHz, CDCl$_3$) δ 152.5, 142.2, 113.8, 97.3, 66.0, 35.2, 31.5, 25.1, 22.5, 14.0 ppm; IR (neat) (ν_{max}, cm^{-1}) 3347, 2930, 2862, 1460, 1048; HRMS (EI) Found 246.0261 [M], C$_{10}$H$_{15}$BrO$_2$ requires 246.0255.

6.6.4 1-(3-Bromofuran-2-yl)Hexan-1-One (14)

Oxalyl chloride (0.9 mL, 11.6 mmol) was dissolved in dichloromethane (40 mL) and brought to –78 °C followed by the addition of DMSO (1.6 mL, 23.3 mmol) and stirred for 5 min. Alcohol **13** (2.6 g, 23.3 mmol) in dichloromethane (5 mL) was added and the reaction mixture was stirred for 15 min at –78 °C followed by the addition of triethylamine (7.4 mL, 52.8 mmol) and stirring was continued for an additional hour. The reaction mixture was brought to room temperature and H$_2$O (40 mL) was added. The mixture was extracted with dichloromethane (3 × 50 mL), washed with H$_2$O (50 mL) and brine (50 mL) and dried over MgSO$_4$. The residue was purified by silica gel column chromatography (pentane/ethyl acetate, 9.5:0.5) to afford **14** (2.1 g, 82%) as a colourless oil. TLC: R$_f$ = 0.44 (pentane/ethyl acetate, 9.5:0.5) ^{1}H NMR (500 MHz, CDCl$_3$) δ 7.47 (d, J = 1.8 Hz, 1H), 6.61 (d, J = 1.8 Hz, 1H), 2.88 (t, J = 7.5 Hz, 2H), 1.74–1.70 (m, 2H), 1.39–1.35 (m, 4H) 0.90 (m, 3H) ppm; ^{13}C NMR (75 MHz, CDCl$_3$) δ 189.5, 148.2, 145.0, 117.3, 106.6, 39.4, 31.4, 23.4, 22.4, 13.9 ppm; IR (neat) (ν_{max}, cm^{-1}) 2956, 2030, 1681, 1550, 1474, 1084; HRMS (ESI) Found 245.0166 [M + H]$^+$, C$_{10}$H$_{14}$BrO$_2$ requires 245.0177.

6.6.5 1-(3-Vinylfuran-2-yl)hexan-1-one (19)

A mixture of Pd(PPh$_3$)$_4$ (70 mg, 0.06 mmol) and LiCl (104 mg, 2.46 mmol) was dissolved in 1,4-dioxane (5 mL) under an atmosphere of N$_2$. Bromide **14** (300 mg, 1.23 mmol) in 1,4-dioxane (1 mL) was added slowly and followed by the addition of tributylvinylstannane (466 μL, 1.59 mmol). The reaction mixture was heated to 120° C using microwave irradiation at 150 W and stirring was continued for 1 h. The reaction mixture was filtered through a small pad of Al$_2$O$_3$ and eluted with EtOAc. The solvent was removed in vacuo and the residue was purified using silica gel chromatography (pentane/Et$_2$O, 98:2) to afford **19** (156 mg, 81% yield) as a colourless oil. TLC: R$_f$ = 0.5 (pentane/Et$_2$O, 9.5:0.5) ^{1}H NMR (500 MHz, CDCl$_3$) δ ppm 7.41 (m, 2H), 6.71 (d, J = 1.5 Hz, 1H), 5.72 (dd, J = 17.7, 1.2 Hz, 1H), 5.42 (dd, J = 11.0, 1.2 Hz, 1H), 2.84 (t, J = 7.5 Hz, 2H), 1.73–1.67 (m, 2H), 1.37–1.33 (m, 4H), 0.90 (m, 3H); ^{13}C NMR (125 MHz, CDCl$_3$) δ ppm 194.4, 144.7, 135.9, 131.3, 129.5, 127.6, 118.6, 42.6, 31.7, 24.6, 22.7, 14.1; IR (neat) (v_{max}, cm^{-1}) 3092, 2957, 2871, 1664, 1426, 1181; HRMS (ES) Found 193.1229 [M + H]$^+$ C$_{12}$H$_{17}$O$_2$ requires 193.1229.

6.6.6 2,3-Dibromo-1-Methyl-1H-Indole (21)

Indole **20** (1.5 g, 12.8 mmol) was dissolved in dry THF (30 mL) in a round bottom flask under nitrogen. The temperature of the flask was lowered to −78° C using dry ice and acetone. *n*-Butyllithium (5.37 mL, 2.5 M in hexanes, 13.44 mmol) was added dropwise. Stirring was continued for 10 min. CO$_2$ was bubbled through the reaction mixture for 10 min by adding small pieces of dry ice directly into the reaction vessel. The reaction mixture was subjected to a vacuum until all bubbling had ceased. THF (30 mL) was added, followed by *t*-BuLi (7.9 mL, 1.6 M in hexanes, 13.44 mmol) to yield a bright yellow colour. Stirring was continued for a further 30 min at –78° C. Dibromotetraflouroethane (3.32 g, 12.8 mmol) in THF (5 mL) was added to the reaction mixture and was allowed to warm to room temperature. The reaction mixture was poured onto water, extracted with diethyl ether (60 mL) and the organic layer was washed with water (2 × 60 mL), dried over MgSO$_4$, filtered and the solvent was removed in vacuo. The residue was treated with DMF (20 mL) and stirred at 0° C. Br$_2$ (0.69 mL, 13.44 mmol) in DMF (10 mL) was added and the reaction mixture was allowed to warm to room temperature. KOH (2.87 g, 51.2 mmol) and MeI (3.188 mL, 51.2 mmol) were added and the reaction mixture was allowed to stir for 16 h. The mixture was poured onto water, extracted with diethyl ether (3 × 60 mL). The organic layers were combined and washed with water (5 × 50 mL), dried over MgSO$_4$ and the solvent was removed in vacuo to yield **21** (2.62 g, 71% yield) as a dark solid, mp 39–40° C; (Lit.[19] mp 38.5–40° C) ^{1}H NMR (DMSO-$d6$, 500 MHz) ppm 7.53 (d, $J = 7.9$ Hz, 1H), 7.41 (d, $J = 7.4$ Hz, 1H), 7.24 (t, $J = 7.4$ Hz, 1H,), 7.16 (t, $J = 7.9$ Hz, 1H), 3.82 (3H, s); ^{13}C NMR (DMSO-$d6$, 125 MHz) ppm 136.1, 126.0, 122.9, 120.9, 117.9, 115.3, 110.8, 91.3, 32.4. IR (CHCl$_3$) (ν_{max}, cm^{-1}) 3582, 3432, 2938, 2360, 2065, 1595, 1461, 1324, 1102. No HRMS data was found for this compound.

6.6.7 1-(3-Bromo-1-Methyl-1H-Indol-2-yl)Hexan-1-ol (23)

A solution of **21** (0.5 g, 1.7 mmol) in dry THF (50 mL) under nitrogen at –78° C was treated dropwise with t-butyllithium (2.25 mL, 3.75 mmol) and was allowed to stir for 5 min. The reaction mixture was treated with hexanal (0.520 mL, 4.25 mmol) in THF (10 mL) and was allowed to warm to room temperature. The reaction mixture was poured onto water and extracted with diethyl ether (2 × 50 mL). The organic layer was washed with water (3 × 50 mL), dried over MgSO$_4$ and solvent was removed in vacuo to yield **23** (290 mg, 53% yield) as a yellow oil after purification using column chromatography on silica gel (CH$_2$Cl$_2$/pentane, 3:2). ^{1}H-NMR (DMSO-d$_6$ 500 MHz) ppm 7.48 (d, $J = 8.3$ Hz,1H), 7.37 (d, $J = 7.9$ Hz, 1H), 7.21 (t, $J = 7.9$ Hz 1H), 7.12 (t, $J = 7.9$ Hz, 1H), 5.63 (s, 1H), 5.05 (m, 1H) 3.89 (s, 3H), 1.9 (m, 2H), 1.75 (m, 2H), 1.4 (m, 2H), 1.2 (m, 2H), 0.8 (m, 3H); ^{13}C-NMR (DMSO-d$_6$ 125 MHz) ppm 139.1, 137.3, 126.2, 122.7, 120.4, 118.5, 110.342, 88.4, 66.4, 40.5, 36.5, 31.8, 31.4, 25.6, 22.5, 14.3; IR (CHCl$_3$) (ν_{max}, cm^{-1}) 3412, 2253, 2128, 1658, 1024; No HRMS data was found for this compound.

General Experimental for Heck Reactions from Tables 6.1 and 6.2.

The palladium source and ligand were dissolved in the appropriate solvent under and atmosphere of N_2. This was followed by the addition of the appropriate base and stirring was continued for the allocated time. The reaction mixtures were continuously analysed by the thin layer chromatography. No products had formed and all starting materials were recovered.

References

1. O' Sullivan TP, Vallin KSA, Shah STA, Fakhry J, Maderna P, Scannell M, Sampaio ALF, Perretti M, Godson C, Guiry PJ (2007) J Med Chem 50:5894
2. Duffy CD, Maderna P, McCarthy C, Loscher CE, Godson C, Guiry PJ (2010) Chem Med Chem 5:517
3. Robert N, Hoarau C, Celanire S, Ribereau P, Godard A, Queguiner G, Marsais F (2005) Tetrahedron 61:4569
4. Wang X-J, Zhang L, Sun X, Xu Y, Krishnamurthy D, Senanayake CH (2005) Org Lett 7:5593
5. Littke AF, Fu GC (2002) Angew Chem Int Ed Engl 41:4176
6. Grushin VV, Alper H (1994) Chem Rev 94:1047
7. Schlosser M, Dinh N (1977) Helv Chim Acta 60:2085
8. Joule JA, Millis K (2010) Heterocyclic chemistry, 5th edn. Wiley, New York
9. Mancuso AJ, Huang S-L, Swern D (1978) J Org Chem 43:2480
10. Scholl M, Ding S, Lee CW, Grubbs RH (1999) Org Lett 1:953
11. Grubbs RH, Chang S (1998) Tetrahedron 54:4413
12. Trnka TM, Grubbs RH (2000) Acc Chem Res 34:18
13. Milstein D, Stille JK (1978) J Am Chem Soc 100:3636
14. Jyotirmayee D, Stellios A, Cossy J (2007) Adv Synth Catal 349:152
15. Li JJ, Gribble GW (2000) Palladium in heterocyclic chemistry, 1st edn. Pergamon, Oxford
16. Bandini M, Eichholzer A (2009) Angew Chem Int Ed Engl 48:9608
17. Bergman J, Venemalm L (1992) J Org Chem 57:2495
18. Bocchi V, Palla G, (1982) Synthesis 12:1096
19. Liu Y, Gribble GW (2002) J Nat Prod 65:748
20. Liu Y, Gribble GW (2002) Tetrahedron Lett 43:7135
21. Katritzky AR, Akutagawa K (1985) Tetrahedron Lett 26:5935

If you have any concerns about our products,
you can contact us on
ProductSafety@springernature.com

In case Publisher is established outside the EU,
the EU authorized representative is:
Springer Nature Customer Service Center GmbH
Europaplatz 3, 69115 Heidelberg, Germany

Printed by Libri Plureos GmbH
in Hamburg, Germany